Blue Ribbon Science Fair Projects

Maxine Haren Iritz

TAB Books

Division of McGraw-Hill

New York San Francisco Washington, D.C. Auckland Bogotá
Caracas Lisbon London Madrid Mexico City Milan
Montreal New Delhi San Juan Singapore
Sydney Tokyo Toronto

pbk 7 8 9 10 11 12 13 14 15 DOC/DOC 9 9 8 7 6
hc 2 3 4 5 6 7 8 9 10 DOH/DOH 9 9 8 7 6 5 4 3 2

Library of Congress Cataloging-in-Publication Data

Iritz, Maxine Haren.
    Blue-ribbon science fair projects / by Maxine Haren Iritz.
       p.   cm.
    Includes bibliographical references and index.
    Summary: Readers learn about the organization and development of school science projects from their beginnings as vague concepts, to the experiment and testing stages, and finally to completion and display.
    ISBN 0-8306-7615-5 (hard)   ISBN 0-8306-3615-3 (pbk.)
    1. Science projects—Juvenile literature. [1. Science projects.]
I. Title.
Q182.3.I75 1991   90-24055
507.8—dc20   CIP
                                      AC

Acquisitions Editor: Kimberly Tabor
Book Editor: April D. Nolan
Director of Production: Katherine G. Brown
Book Design: Jaclyn J. Boone
Cover Photography: Dennis Brooks

Contents

Acknowledgments   **v**

Introduction   **vii**

1 In the Beginning . . .   **1**
Choosing a topic   1
Topics best avoided   6

2 Earth Day   **9**
The good earth   10
It's raining, it's pouring   18
Helpful hints   25

3 Around the House   **31**
Jolly Orville   31
Green thumb   37
Rusty Nails   41
Helpful hints   42

4 Computer Science   **47**
Do you speak BASIC?   47

5 All About Us   **63**
Seeing is believing   63
Where there's smoke . . .   69
Helpful hints   76

6 Fun and Games   **79**

Home field advantage   79
That's the way the ball bounces   86
Helpful hints   90

7 Science Project Organizer   **95**

Appendix   ISEF Project Categories   **109**

Glossary   **113**

Bibliography   **115**

Index   **118**

To:

Aaron
Brian
Catherine
Daryl
Davis
Jennifer
LeMar
Matthew
Vy
Yolanda

Thank you.

Introduction

*F*or many of you, a science project is the largest, most complex piece of work you've ever attempted. Although the individual parts of the project might not be difficult, to complete the whole task you'll need to use a combination of skills, some of which you never knew you had. Many of these skills or qualities are obvious ones, such as perseverance, organization, and promptness. (Don't worry, we won't include politeness, although it's not a bad idea when the time comes to talk to science fair judges.) Other attributes, such as creativity and flexibility, are a little harder to pin down, but these are qualities that will be very helpful. If you don't start the project with these abilities, you'll probably have them by the time you finish.

While you're working on your project, you'll develop a better knowledge of yourself. Because science projects involve so many different stages, you'll learn whether you love research, are afraid of handling chemicals, are excited when the statistics come out just the way you predicted, or enjoy creating an attractive display.

You'll also get a strong feeling for the way you like to work. Will you want to work precisely on Thursday the 23rd between 2 and 4 p.m., or any time between the first day of school and President's weekend? Some people like a firm, structured schedule, with every task broken down to its smallest element and assigned to a specific time slot. Others are well disciplined and can pace themselves and set their own goals. There are those, however, who like to live dangerously. They perform well only when they have a fast-approaching deadline, with more work than they can reasonably handle (often to the frustration and panic of their parents and teachers). Some people prefer to do one task at a time, and others can work on many things at once. Structured, flexible, organized, or chaotic—for most of you, a science project is the first opportunity to learn how you work best.

This book presents fully developed science projects in a variety of categories. These projects are grouped by theme rather than by category, to illustrate the many ways you can get science project ideas. Incidentally, all the projects shown are actual projects that students developed. As you will see, some of these projects are fairly easy to do, while others are quite complex. Not all of these projects are prize winners, but they are all top-notch projects done by students just like you.

You'll see every step in the development of these projects—from vague concept to science fair display. For each project, you'll learn how students found their topics, conducted project research, and developed the question and hypothesis.

You'll see how they got organized to begin their experiments and get some ideas on how they did their research. (The bibliographies for some students' projects are located in the back of this book.) You will learn how the students bought, borrowed or built the supplies and materials they would need. By analyzing how these students completed their projects, you'll learn how to conduct an experiment, or write and test a computer program. You'll find out how each student reached his or her conclusion and how each experiment, device, or program was displayed to its best advantage. Finally, we'll include some helpful hints for working on your own projects. To help you get started on your project, we've included a quick step-by-step science project organizer in Chapter 7.

Whether or not you decide to focus on a career in science, computers, or engineering, you'll find that the experiences you have and the lessons you learn while doing a science project have lasting benefits. One young man who worked on several science projects in junior and senior high school faced a large sociology project during his freshman year in college. Most of his classmates were in a panic, but he was quite calm. "It's just another science project," he said.

So look at it as an adventure—a voyage of discovery.

Chapter 1

In the beginning...

*A*ll science projects begin with two basic elements—an idea or an inspiration, and the ability and enthusiasm to complete the project. For many of you, a science project starts even before that, when your science teacher announces that you must do a project if you have any hopes of passing the class. (Don't try to get out of it; the points are probably rigged against you). Once you have your idea, you'll need the information, equipment, assistance, and perseverance (or perspiration, if you'd rather call it that) to complete the experiment, program, or device.

CHOOSING A TOPIC

Probably the hardest part of any project is deciding on a topic. You want something that's not too simple, but not so complex that it's difficult to handle. You don't want to work on a project idea that's been entered in every science fair since the turn of the century, but you can't pick something so obscure that no information exists outside an underground, top-secret lab.

One of the best ways to find a good idea is to concentrate on things that interest you. Remember, your science project is something that you'll be living with for the better part of the school year. If you like your topic, it's a good bet that you'll enjoy the project, too. If you enjoy the project, the odds are that you'll finish it successfully.

Suppose you're interested in sports, music, cars, food, and boys (or girls). So is everyone else. Of course, that brings you to the next problem—finding something unique. You might have a particular hobby or interest that will give you an idea. Do you play chess, water ski, fish, or sew? You might find an idea in one of those pursuits. Figure 1-1 through 1-6 show some interesting and unusual project ideas.

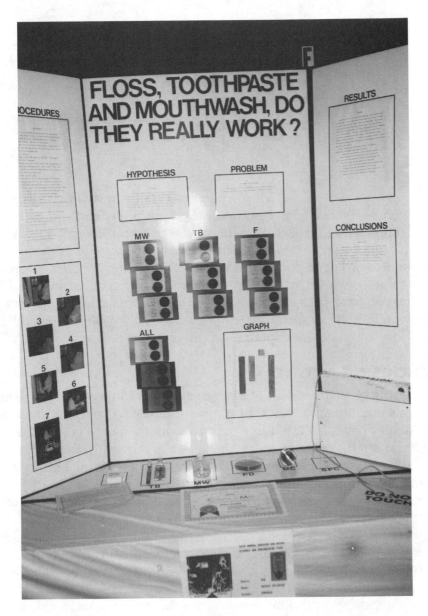

1-1 Backboard.

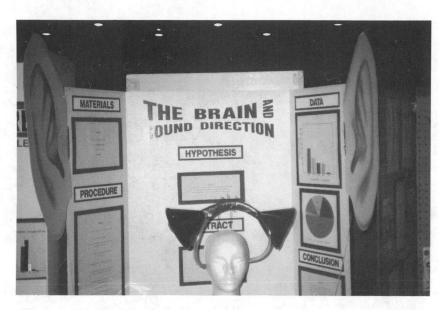

1-2 Backboard.

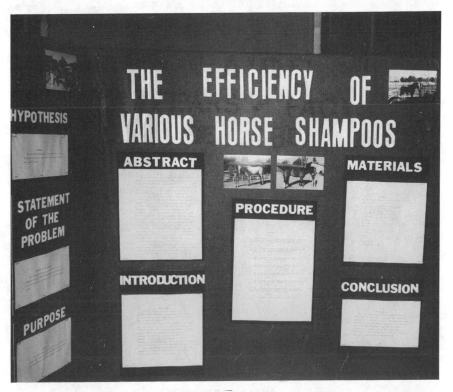

1-3 Backboard.

1-4 Backboard.

1-5 Backboard.

1-6 Backboard.

One way to find out if your idea is original is to look at programs from previous science fairs to see what's been done (and overdone). Your teachers will probably be glad to tell you what projects they never want to see again in this lifetime. If you're lucky, you'll get a list of "don'ts" before you've fallen in love with one of these ideas.

If you're interested in something you're currently studying in class, that's an excellent place to start, since you'll already know something about the topic, and your textbook may contain additional sources of information. The trick here, though, is to put a new twist on the topic.

Another place you can look for a project idea is in your family. "But that's cheating," you might think, "doing a project on motor oils when my aunt and uncle own a garage." Don't believe it for a minute. For most projects, you'll need some sort of assistance, and there's no place like home! In fact, some students have developed excellent projects because they became interested in a parent's profession or hobby. As long as you don't have someone else doing your work, there's nothing wrong with getting your idea or inspiration from someone you know.

Current events are also excellent sources of project ideas. In such areas as ecology, product safety, and artificial ingredients in foods, today's news item can be tomorrow's science project. Product claims can also suggest experiments. "New and improved," screams the label. Really? How? Perhaps a science project will show you.

Regardless of where you've found your project idea, you'll need to do a reality (or sanity) check to make sure you have the time and ability to complete the project. Your teacher can help you take a realistic look at your project idea and decide if it's a solid concept that's right for you.

Many students think doing a project is a signal to take on some of society's great problems. It's important to come back to reality because it's highly unlikely that you and your science project will find a cure for AIDS or end world hunger. However, you could do a project that addresses some important concerns on a smaller scale that you can handle. Also, the work that you do now might be the beginning of greater efforts in your future. The main reasons for not biting off more than you can chew right now are to keep you from getting discouraged and to make sure that you can actually do the project in the time you have.

TOPICS BEST AVOIDED

Besides the list of "tired old projects" that you'll no doubt get from your teacher, here are a few more no-no's. Although no category is completely off-limits, in these areas, proceed at your own risk.

Projects involving live vertebrates are becoming more difficult to enter into science fairs than in previous years. Animal rights groups are becoming more vocal, even when the animals are used in ongoing research designed to find cures for human diseases. For this reason, animal rights activists will probably be very critical when research animals are used only to benefit a student's grade.

Because of this controversy, science teachers and judges are extremely careful when reviewing a project that involves animals. If any of the experimental animals die, judges might determine that the deaths occurred as a result of the experiment and will not allow your work into a science fair. For example, although mice commonly attack each other, animal activists might assume they attacked because of overcrowding. Unless you have a professional advisor who can certify that there is no mistreatment of the animals at any time during your project, find another project idea. (Your family will probably be pleased, too!)

You'll also need to be careful if you plan to use human subjects. Any experiment that might potentially cause physical or psychological damage must be done under supervision. Some topics are obviously unacceptable, such as depriving people of sleep or nourishment. But there are many projects you can do, such as the two projects presented in Chapter 5. As you will see, these experiments required the subjects to perform two very normal functions—breathing and seeing.

At times, it is difficult to determine whether an experiment could cause harm. A recent controversy involved a project where adults were given alcoholic beverages and then tested on a Nintendo game. (Incidentally, the more they drank, the worse they scored.) Although the project won at a school science fair, a screening committee later disqualified it. The judges determined that the tests, which were done without supervision, could have caused physical or psychological harm. MADD (Mothers Against Drunk Driving) disagreed with the ruling, but the judges stood firm, and the project remained disqualified.

So before you begin a project that uses human subjects, check with your teacher or, if possible, discuss it with your local screening committee. If you're committed to the idea but need professional supervision, try to find someone in the field who will oversee your experiment. Still in doubt? Find another topic.

In spite of these few warnings, there's a whole world of ideas just looking for a home in your science fair project. The next few chapters take a look at several projects, from the easy to the complex, from the commonplace to the unusual, that students just like you have developed. You'll see that regardless of your interests or expertise, there's something just waiting for you. Let's go!

Chapter **2**

Earth day

With so many important issues in the
world today, many of you are probably concerned about the future
of our world. Some of these issues—such as war, poverty, and hunger—might not immediately give you a science project idea. However, the future of our planet and its resources can inspire some
interesting science projects.

Projects that deal with preserving and protecting the earth can
fall into almost any category. Engineering projects might find a
more efficient way to use energy; chemistry can test methods of
purifying our resources; botany might find ways to grow more
food; zoology can determine the effects of various substances on
animal life. However, most projects that deal with ecological issues
fall into the categories of Earth and Space Sciences, or Environmental Sciences.

If you are interested in these areas, a science project idea is no
further away than your radio, television, or newspaper. However,
because most of these questions are very broad and often complicated, you will probably need to narrow down the idea to something you can do in a few months.

The next two projects deal with important issues for our
time—retaining our soil and preventing the polluting effects of acid
rain. In both cases, students found a small, manageable part of a
broader topic. Brian Berning's idea to do an experiment involving
erosion was based on a project he had previously seen. He
expanded the sample size to create a more comprehensive experiment. Catherine Davis, on the other hand, could not adequately

cover the topic of acid rain in one semester. Therefore, she limited her project to San Diego, using rainwater samples from five locations throughout the county.

THE GOOD EARTH

Brian Berning found his topic when he saw a project about erosion at another school's science fair. He decided that the following year, he would also test soil erosion using several soil types. Over the summer vacation, he and his family visited relatives in other parts of the country. While traveling, he collected samples of Ohio farm soil and Virginia clay and brought them home—despite protests from members of his family, who objected to using luggage space for dirt. In September, Brian gathered three samples from San Diego—soil, topsoil, and sod. These five samples formed the basis of his science project in the Earth and Space Sciences category.

For his background research, Brian used a branch library and the central library in San Diego. His bibliography is included in the back of this book. As he read, Brian took notes on index cards. He then separated his index cards into several categories, representing the paragraphs in the research paper. Once the material was organized, Brian wrote the paper using word processing software on a computer. He then developed the project question and hypothesis, shown in FIG. 2-1. The experimental variables and controls and the five experimental groups are shown in FIGS. 2-2 and 2-3.

Question	Hypothesis
Do some soils erode faster than others?	Sandy soil erodes faster than sod.

2-1 Question and hypothesis; *Erosion and Its Effects*.

To prepare for the experiment, Brian's father helped him build a rack that he could use to conduct and display the project. This rack, constructed of wood and nails, is shown in FIG. 2-4. The complete materials list for the project is shown in FIG. 2-5.

Once all the materials were assembled, Brian set up the experiment by filling five trays with the soil samples and weighing each one on a kitchen scale. He then arranged the samples, the run-off trays, and the "watering" cans on the rack, as shown in FIG. 2-6.

Experimental Groups
1. California Sod
2. California Soil
3. California Topsoil
4. Ohio Farm Soil
5. Virginia Clay

2-2 Variables and controls; *Erosion and Its Effects*.

Variables	Controls

Experimental
   Soil samples
 	
Measured
   Weight of soil
   samples
	Amount and rate of water
	given for each sample

2-3 Experimental and control groups; *Erosion and Its Effects*.

2-4 Rack used to display experiment for *Erosion and Its Effects*.

Materials

1. Wooden rack

2. Five soil samples, Ohio farm soil, Virginia clay, California soil, California topsoil, and California sod.

3. Five interlocking Rubbermaid storage trays, with small drainage holes hammered in, to hold the soil samples.

4. Five interlocking Rubbermaid storage trays to catch the runoff soil and water.

5. Five peanut cans, with nails hammered into the bottoms of the cans. This allowed gradual watering of the soil samples, rather than having all the water absorbed at once.

6. Kitchen scale.

2-5 Materials used for *Erosion and Its Effects.*

2-6 Experiment samples displayed on home-made rack.

Brian conducted his entire experiment at home, over a period of two weeks. During the experimental period, Brian tested his samples every day. In the morning, he weighed each specimen and recorded the measurements in his log. An example of what his daily log might have looked like is shown in FIG. 2-7. He then

Amount of Water - 4 cups

12-7-89

1. 1 pound 14.5 ounces   Ca. Topsoil
2. 2 pound 4 ounces   Ca. Soil
3. 1 pound 14 ounces   Virginia Clay
4. 1 pound 14.5 ounces   Ohio Soil
5. 2 pounds 6 ounces   Ca. Sod

12-8-89

1. 1 pound 14.5 ounces Ca. Topsoil
2. 2 pounds 4.5 ounces Ca. Soil
3. 1 pound 14 ounces Virginia Clay
4. 1 pound 14.5 ounces Ohio Soil
5. 2 pounds 6 ounces Ca. Sod

12-9-89

1. 1 pound 14.5 ounces Ca. Topsoil
2. 2 pound 4 ounces Ca. Soil
3. 1 pound 14 ounces Virginia Clay
4. 1 pound 14.5 ounces Ohio Soil
5. 2 pound 6 ounces Ca. Sod

2-7 Example of experimental log.

poured 2 cups of water into the cans above each soil sample and allowed the water to drip into the soil. Each night, he watered the soil again and allowed it to dry out overnight.

While conducting the experiment, Brian observed that the erosion occurred in different patterns in the various soil samples. For example, the California topsoil showed a pattern of gullied erosion, with a lot of dirt running off. In the Virginia clay and the Ohio farm soil, however, the dirt spread out evenly when it was watered. The daily experimental procedures are summarized in FIG. 2-8.

Procedures

1. Every morning, weighed each soil sample on the kitchen scale and recorded the measurements in the project log.

2. Every morning and evening, poured two cups of water into the cans above each soil sample, and allowed the water to drip into the soil.

3. Each night, allowed samples to dry out.

2-8 Procedures for *Erosion and Its Effects*.

At the end of the two week period, Brian summarized the recorded data, as shown in FIG. 2-9. He used this data to graph the weights of each soil sample. Figures 2-10 through 2-14 show the graphs for California topsoil, California soil, California sod, Ohio farm soil, and Virginia clay.

The results indicated that California topsoil, which is sandy soil, eroded the most. California sod eroded the least. Therefore, the results of Brian's experiment proved his hypothesis.

When assembling his backboard, Brian used Print Shop on his Apple IIe to create the titles and other printed materials. In front of the backboard, Brian displayed the actual rack, together with the experimental samples. The complete display for the project, entitled "Erosion and Its Effects," is shown in FIG. 2-15.

Brian felt satisfied with his project—so satisfied, in fact, that next time, he will try a completely different topic.

	11/25	11/26	11/27	11/28	11/29	11/30	12/1	12/2	12/3	12/4
CA Topsoil	31.0	36.0	36.0	33.5	37.0	32.0	31.5	31.0	31.0	31.0
CA Soil	30.0	36.5	36.5	36.5	36.0	36.0	36.0	36.5	36.0	36.0
VA Clay	28.5	30.5	30.5	29.5	30.0	30.0	30.0	29.5	29.5	29.5
OH Farm Soil	30.0	30.5	30.5	29.5	30.0	30.0	30.5	30.5	30.0	30.5
CA Sod	35.0	36.0	36.0	36.5	36.5	36.5	37.5	37.0	37.0	37.5

2-9 Summarized recorded data.

CALIFORNIA TOPSOIL

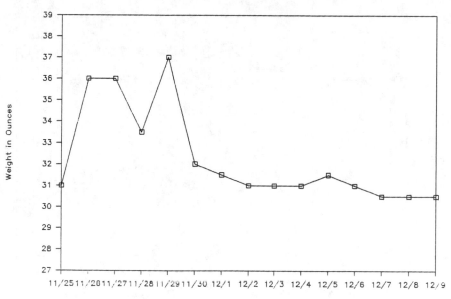

2-10 Graph of California topsoil.

CALIFORNIA SOIL

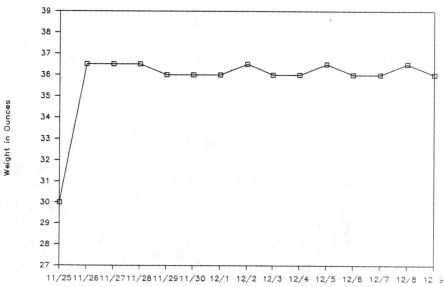

2-11 Graph of California soil.

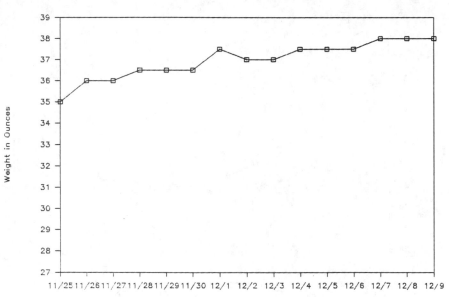

2-12 Graph of California sod.

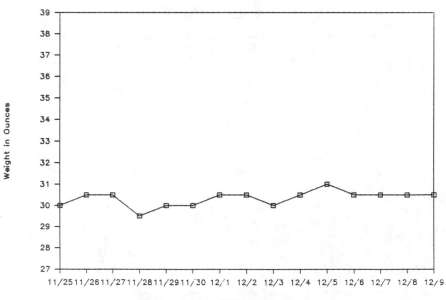

2-13 Graph of Ohio farm soil.

VIRGINIA CLAY

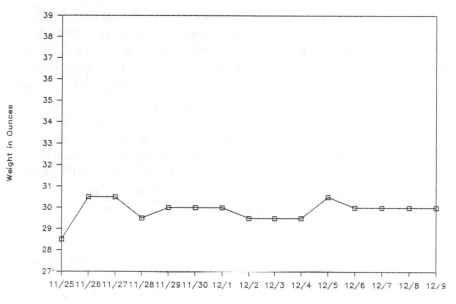

2-14 Graph of Virginia clay.

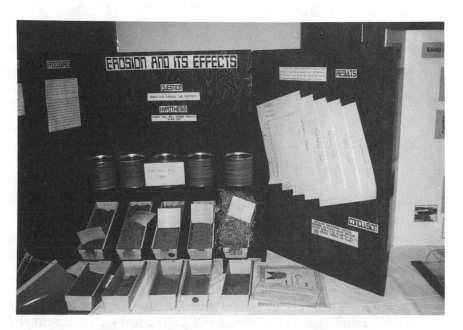

2-15 Backboard display for *Erosion and Its Effects*.

IT'S RAINING, IT'S POURING

Catherine Davis's project, "San Diego County's Acid Rain," began as a topic that was too broad and too complex. However, with some good organization and professional help, the project turned out to be quite successful. This project also had to overcome a problem that many students confront—a lack of supplies. In this case, a drought produced a shortage of rain. Although in the end there was sufficient rainfall for Catherine to finish her project, it is always a good idea to have an alternate plan in mind.

Catherine was interested in the issues of ecology and pollution, especially those involving water. She is interested in these fields because her grandfather is an instructor in Environmental Studies at the University of San Diego (USD), and she wanted to do a project in the Environmental Sciences category.

Catherine conducted her research in four public libraries, as well as the USD library. She used several sources of information (See her bibliography at the end of this book). Catherine took notes on index cards, which she then grouped according to topic. She did not use a written outline, and wrote the background research paper using a computer. Catherine then developed the project question and hypothesis, shown in FIG. 2-16.

Question	Hypothesis
Is San Diego County endangered by acid rain?	The acidity of rain in San Diego County is porportional to the distance from the ocean.

2-16 Question and hypothesis *San Diego Acid Rain*.

Catherine's experiment involved collecting samples of rainfall from five San Diego county locations—Clairemont, El Cajon, Hillcrest, La Jolla, and Tierrasanta—testing the acidity of the samples, and then statistically analyzing the results. To conduct the experiment, she used materials which cost less than $15.00 total (FIG. 2-17).

At each location, two people volunteered to collect samples for her. Catherine gave each of her assistants a test kit, consisting of two jars and mailing labels. She also gave each volunteer an instruction sheet, shown in FIG. 2-18.

The samples were collected between November 28, 1989, and

Materials
1. Markson temperature pH meter
2. Thermolyne Midget Stir Plate
3. 2 dozen one pint Ball canning jars and lids
4. 1 set standard mailing labels
5. 10 plastic grocery bags
6. 4.01 and 6.68 pH buffer solution
7. Small squirt bottle of distilled Water
8. One large Pyrex bowl
9. One test tube
10. Two 8 ounce beakers

2-17 List of materials for *San Diego Acid Rain* project.

Directions for collecting rain:

1. **Set out a large clean glass bowl when you suspect it will rain.**
2. **The next morning, take the rain you have collected and pour it into one of the jars.**
3. **Mark the label with the date and the location in San Diego where the rain was collected.**
4. **I will arrange to pick up the jars from you as soon as possible.**

2-18 Instructions for collection of rainfall.

January 17, 1990. To make sure that she had enough samples, especially during the holiday season, Catherine sometimes had to remind her volunteers to collect, but everyone seemed eager to help. During November and December, another problem was a lack of rainfall, and Catherine worried that she wouldn't have enough samples to finish the project. However, there were a few storms in January, some lasting several days and yielding a number of samples.

As soon as possible after each rainfall, Catherine picked up her samples. She originally planned to test the rain water at the USD laboratory, but had problems with the equipment, so she did all the testing at home. The testing consisted of two phases—calibrating the meter, and testing the samples.

First, Catherine calibrated the pH meter. She mixed buffer material with four ounces of distilled water in an 8-ounce beaker, using a Thermolyne Midget Stir plate. Then she cleaned the electrode by spraying it with distilled water, as shown in FIG. 2-19. She stirred the needle in the buffer solution, as shown in FIG. 2-20, and calibrated it to pH 6.86. She cleaned the electrode again, and then calibrated the meter to a temperature of 4.01.

2-19 Cleaning the electrode.

Once the meter was calibrated, Catherine was ready to test her samples. She again cleaned the electrode and placed it in the jar of rain water, as shown in FIG. 2-21. She gently shook the water until the needle stopped, took the reading, and recorded it in a daily log. A summary of the experimental procedures is outlined in FIG. 2-22.

When all samples were tested, Catherine entered her findings into the computer and listed them by location and date, as shown in FIGS. 2-23 and 2-24. This data formed the basis of the statistical analysis she needed. Catherine had the benefit of some expert help here, too; her mother teaches mathematics and statistics.

Catherine computed descriptive statistics for each date and location tested. The statistics by location are shown in FIG. 2-25. By

2-20   Calibrating the pH meter.

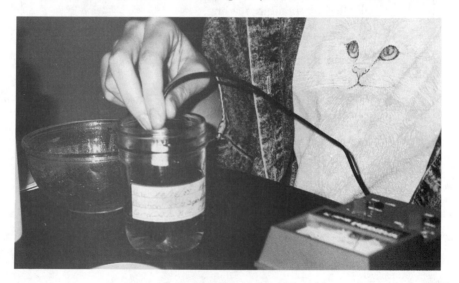

2-21   Testing the rainfall sample.

using a t-test, she also compared the results against the average acidity for rainwater, which is between 5.0 and 6.5. The t-tests showed the statistical difference of each location from the norm, which is 5.60 (FIG. 2-26). Catherine tested each location against the other four locations for statistical significance, as shown in FIG. 2-27. Then she tested each date against the other dates, as shown in FIG. 2-28.

Procedures

1. Calibrated pH meter as follows:

   a. Mixed buffer solution with 4 ounces of distilled water in an 8 ounce beaker, using a Thermolyne Midget Stir Plate.

   b. Cleaned electrode by spraying it with distilled water from squeeze bottle.

   c. Placed electrode in 6.68 buffer solution and calibrated the meter to pH 6.68.

      Note: The meter readings range from 2 through 12. Readings below 7 are acidic, and readings above 7 are alkaline. A reading of 7 is neutral.

   d. Cleaned electrode by spraying it with distilled water from squeeze bottle.

   e. Calibrated electrode using 4.01 buffer solution, and adjusted temperature reading.

2. Tested each sample, as follows:

   a. Cleaned electrode by spraying it with distilled water from squeeze bottle.

   b. Placed electrode in a jar of rain water.

   c. Gently shook water until the needle on the pH meter stabilized, and recorded the reading, to the nearest hundredth, in log.

      Note: If there was insufficient rain water in the jar, poured the sample into the test tube before testing.

2-22 Procedures for *San Diego Acid Rain*.

The results showed that San Diego county rainfall generally falls within the normally accepted range for acidity. The statistics also indicated that the rainfall from Tierrasanta and El Cajon, which are farthest from the ocean, was less acidic than the other rainfall samples. The rainfall from Hillcrest, which was the most geographically central, was the most acidic. Catherine determined that this condition might exist because Hillcrest is close to downtown San Diego, an area with a great deal of traffic and smog.

An unexpected conclusion shows that the rainfall on the day

Data

pH Readings by Location

La Jolla	Clairmont	Hillcrest	Tierra Santa	El Cajon
6.00	5.80	4.30	6.25	6.45
5.50	3.90	5.45	5.85	5.65
4.35	3.95	5.75	5.95	5.65
3.80	4.75	5.15	4.90	5.85
5.82	5.65	5.95	5.35	5.85
5.25	5.55	5.65	5.95	6.65
5.95	5.99	5.98	6.25	
4.65	6.05		5.72	
5.85	6.15		6.10	
6.01	6.50		6.60	
6.15	6.35		6.15	
6.25	6.20			
4.65	6.20			
5.70				
5.85				
5.70				
5.65				
6.00				
5.85				

2-23 Summarized recordings.

after the clouds were seeded, December 28, showed acidity of 4.97, which is above the normal range. This result is shown in FIG. 2-29.

Catherine constructed her backboard of painted cardboard (a really inexpensive solution—she used leftover wall paint). To title the panels, she used 2-inch and 3-inch vinyl letters. Besides her graphs and charts, she featured a map with the test locations color-coded to the descriptive statistics for each location, shown in rain-drop cutouts. The backboard is shown in FIG. 2-30.

Catherine enjoyed the project, and anticipates doing another science project involving acid rain. As a result of her unexpected conclusion, next year she will study the effects of cloud seeding on

pH Readings by Date

11/28	12/28	1/2	1/7	1/12 -1/13	1/13	1/13 -1/14
6.00	5.85	5.65	5.75	5.65	4.65	6.15
5.80	5.95	5.55	5.15	5.95	5.85	5.75
6.25	5.50		5.95	5.99	6.05	6.01
6.45	4.35		4.90	5.55	5.85	5.85
	3.80		5.35		6.25	5.70
	4.30		5.65			
	3.90					
	3.95					
	4.75					
	5.82					

1/14	1/14 -1/15	1/15 -1/16	1/16 -1/17
5.95	6.35	4.65	6.20
6.65	6.25	5.98	6.15
6.15	6.60	5.70	6.20
6.10	5.65		5.65
6.50	5.85		6.00
			5.85

2-24 Summarized recordings.

acid rain. She would also like to eliminate the stress of calculating the statistics at the last minute, but as long as she has to wait for the rain in Southern California, that problem may be unavoidable!

HELPFUL HINTS

☐ While doing research, organize your facts, either by using index cards or by identifying each fact by category. It will make writing your paper easier.

☐ Do your experiment yourself, but take advantage of any resources that are available to you, either for information, assistance, or equipment.

☐ If you're working with a large topic, narrow it down to an area that's more manageable. If your original idea is too limited, use more samples or trials to make the results more conclusive.

☐ Your experimental log can be handwritten, as long as you can read it when it's time to compile your results.

𝔇𝔢𝔰𝔠𝔯𝔦𝔭𝔱𝔦𝔳𝔢 𝔇𝔞𝔱𝔞

Descriptive Statistics by Location

La Jolla		Observations:	19
Minimum: 3.800		Maximum:	6.250
Range: 2.450		Median:	5.820
Mean: 5.525		Standard Error:	0.155
Variance:		0.457	
Standard Deviation:		0.676	

Clairmont		Observations:	13
Minimum: 3.900		Maximum:	6.500
Range: 2.600		Median:	5.990
Mean: 5.618		Standard Error:	0.242
Variance:		0.760	
Standard Deviation:		0.872	

Hillcrest		Observations:	8
Minimum: 4.300		Maximum:	5.980
Range: 1.680		Median:	5.600
Mean: 5.472		Standard Error:	0.193
Variance:		0.297	
Standard Deviation:		0.545	

2-25 Descriptive statistics by location.

t-test for a Difference from the Norm of 5.60

t - test of Location Against the Norm (5.60)

La Jolla		Population
Mean:	5.525	5.600
Std. Deviation:	0.676	
Observations:	19	

t-statistic:	-0.482	Hypothesis:
Degrees of Freedom:	18	Ho: $\mu1 = \mu2$
Significance:	0.636	Ha: $\mu1 \neq \mu2$

Clairmont		Population
Mean:	5.618	5.600
Std. Deviation:	0.872	
Observations:	13	

t-statistic:	0.076	Hypothesis:
Degrees of Freedom:	12	Ho: $\mu1 = \mu2$
Significance:	0.940	Ha: $\mu1 \neq \mu2$

Hillcrest		Population
Mean:	5.472	5.600
Std. Deviation:	0.545	
Observations:	8	

t-statistic:	-0.662	Hypothesis:
Degrees of Freedom:	7	Ho: $\mu1 = \mu2$
Significance:	0.529	Ha: $\mu1 \neq \mu2$

2-26 t-Test—Locations vs. the norm.

t-test for a Difference Between Locations

	La Jolla	Clairmont
Mean:	5.525	5.618
Std. Deviation:	0.676	0.872
Observations:	19	13

t-statistic:	-0.340	Hypothesis:
Degrees of Freedom:	30	Ho: $\mu1 = \mu2$
Significance:	0.736	Ha: $\mu1 \neq \mu2$

	La Jolla	Tierra Santa
Mean:	5.525	5.915
Std. Deviation:	0.676	0.467
Observations:	19	11

t-statistic:	-1.689	Hypothesis:
Degrees of Freedom:	28	Ho: $\mu1 = \mu2$
Significance:	0.102	Ha: $\mu1 \neq \mu2$

	La Jolla	El Cajon
Mean:	5.525	6.017
Std. Deviation:	0.676	0.427
Observations:	19	6

t-statistic:	-1.664	Hypothesis:
Degrees of Freedom:	23	Ho: $\mu1 = \mu2$
Significance:	0.110	Ha: $\mu1 \neq \mu2$

2-27 t-Test—Location vs. location.

t-test for a Difference Between Dates

	11/28	12/28
Mean:	6.125	4.976
Std. Deviation:	0.284	0.827
Observations:	4	13

t-statistic:	2.677	Hypothesis:
Degrees of Freedom:	15	Ho: $\mu1 = \mu2$
Significance:	0.017	Ha: $\mu1 \neq \mu2$

	11/28	1/2
Mean:	6.125	5.600
Std. Deviation:	0.284	0.071
Observations:	4	2

t-statistic:	2.437	Hypothesis:
Degrees of Freedom:	4	Ho: $\mu1 = \mu2$
Significance:	0.071	Ha: $\mu1 \neq \mu2$

	11/28	1/7
Mean:	6.125	5.458
Std. Deviation:	0.284	0.395
Observations:	4	6

t-statistic:	2.886	Hypothesis:
Degrees of Freedom:	8	Ho: $\mu1 = \mu2$
Significance:	0.020	Ha: $\mu1 \neq \mu2$

2-28   t-Test—Date vs. date.

Results

After running the data I discovered that the rain water for 12 /28 was the only rain water that is outside the range for normal rain water, which is 5.00 - 6.50.

1 2 / 2 8	Observations: 1 3
Minimum: 3.800	Maximum: 5.950
Range: 2.150	Median: 5.250
Mean: 4.976	Standard Error: 0.229
Variance:	0.684
Standard Deviation:	0.827

Most of the other dates were more alkaline than the regular 5.60, although none of them were outside the range. These dates were 11/28, 1/13 - 1/14, 1/14, 1/14 - 1/15, and 1/16 - 1/17.

The locations were all in this range also. Although Tierra Santa and El Cajon were statistically more alkaline than the norm of 5.60, they both still fell within the range of normal rain water.

Tierra Santa		Population
Mean:	5.915	5.600
Std. Deviation:	0.467	
Observations:	11	
t-statistic:	2.242	Hypothesis:
Degrees of Freedom:	10	Ho: $\mu1 = \mu2$
Significance:	0.049	Ha: $\mu1 \neq \mu2$

2-29 Statistics for one day, December 28, 1989.

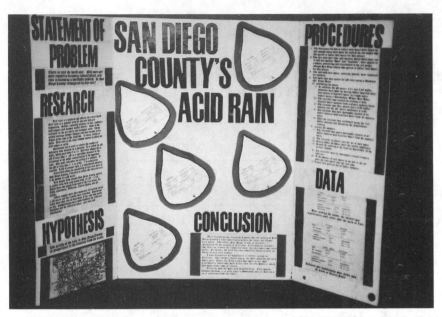

2-30 Backboard display for *San Diego Acid Rain*.

Chapter **3**

Around the house

Some good project ideas can come from the simplest, most obvious sources. Just looking around the house, examining the food that you eat, the clothes that you wear, and the tools, products, and appliances that you use can give you an idea. These projects can fall into almost every category, from human behavior to chemistry to engineering. The following projects will demonstrate that you do not need to turn to complicated or obscure topics to find your science project. Some excellent project ideas often come from some common, every day items.

JOLLY ORVILLE

Nguyen Vy was rather unenthusiastic about a science project and wanted something that related to everyday life. She came up with a list of ideas in several categories, but because one of Vy's main interests is cooking, she really wanted a topic in that field, so Vy decided to test how well popcorn popped under different conditions.

Her teacher conducted several brain-storming sessions, both with the entire class and with individual students. As a result of these sessions, Vy planned to test room-temperature popcorn and frozen popcorn to see which popped better. Originally, she thought she would freeze the popcorn for 20 minutes, but, together with her teacher, decided that a longer freezing time would be better. They determined that the popcorn should be frozen for three days.

Vy conducted most of her research in the school library, using the encyclopedias. She also found an Orville Redenbacher recipe book, which was helpful. Vy's bibliography appears at the end of this book.

While doing research, Vy took notes on sheets of paper. When it was time to write the research paper, she looked for related ideas and numbered them, so that she could easily arrange them into paragraphs. Vy decided to produce a handwritten draft of the paper, edit it, and then type a final copy.

The project question and hypothesis are shown in FIG. 3-1, and the variables and controls are shown in FIG. 3-2. This project used two experimental groups and no control groups, as shown in FIG. 3-3.

Question	Hypothesis
Does frozen popcorn pop better?	Popcorn at room temperature pops better.

3-1   Question and hypothesis; *Cool It, Orville Redenbacher.*

Variables	Controls
Experimental Temperature of popcorn **Measured** Number of popped kernels Number of unpopped kernels	• Popping time • Popping method • Type of popcorn

3-2   Variables and controls; *Cool It, Orville Redenbacher.*

Experimental Groups	Control Group
• Frozen popcorn	• Room temperature popcorn

3-3   Experimental and control groups; *Cool It, Orville Redenbacher.*

Because of the number of batches tested, this experiment was time-consuming. However, Vy completed her project with a minimum of materials and expense, as shown in FIG. 3-4. She was able to conduct the entire experiment at home, and needed help only from her mother as a "safety inspector."

Materials
1. Jolly time popcorn
2. Cooking pot
3. Bowls
4. Paper bags

3-4 Materials list; *Cool It, Orville Redenbacher.*

To ensure that she did enough testing, Vy conducted her experiment 12 times with frozen popcorn and 12 times with room-temperature popcorn. For each trial with frozen popcorn, she took 200 kernels and froze them for three days in an uncovered bowl. Then she placed the kernels in a pot and popped them without oil for two minutes. She quickly poured the popcorn into a bowl, and transferred the popped kernels to a bag. Vy counted the unpopped kernels left in the bowl, subtracted the number from the original 200 kernels, and recorded the number on her experimental log.

Vy used exactly the same procedure with the room-temperature popcorn. She conducted the 24 trials over a period of three weeks. Any time she was interrupted or an experiment failed due to spillage or other circumstances, she discarded the sample and started again. Vy's procedures are summarized in FIG. 3-5.

When all testing was complete, Vy totaled the numbers for popped and unpopped kernels in each category and calculated the average and mean number of kernels popped. She then graphed the results for frozen popcorn, FIG. 3-6, and room-temperature popcorn, FIG. 3-7. After analyzing the data, the experiments showed that popcorn kept at room temperature popped better by an average of $3^3/4$ kernels out of 200.

Vy's backboard consisted of three equal-sized panels, using paper attached to plywood stretcher frames. She originally drew up a plan of what belonged on each panel, but when it was time to create the backboard, she simply attached the material, including

Procedures
1. Popped 200 kernels of popcorn in a pot without oil for 2 minutes
2. Poured popcorn into a bowl
3. Transferred popped kernels into a bag
4. Counted unpopped kernels left in bowl
5. Subtracted number of unpopped kernels from 200 and recorded the resulting number of popped kernels in the experimental log

3-5 Procedures list; *Cool It, Orville Redenbacher.*

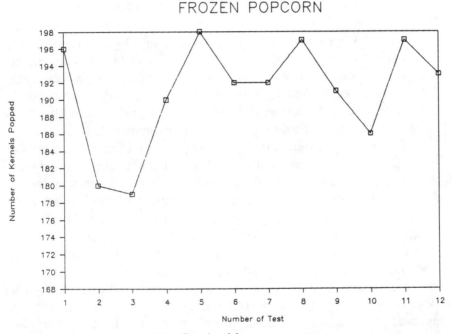

3-6 Graph of frozen popcorn.

bags of popcorn, where it fit best. Partly inspired by the recipe book she found in the library, she entitled her project, entered in the Engineering category, "Cool It, Orville Redenbacher." In a way, the backboard was the most difficult part of the project because she had to stay up late to finish it. Views of Vy's backboard are illustrated in FIGS. 3-8, 3-9, and 3-10.

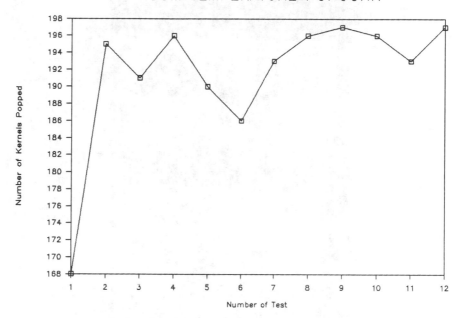

3-7 Graph of room-temperature popcorn.

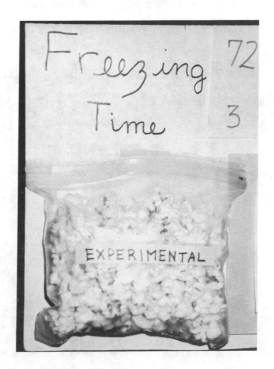

3-8 Creative backboard display for *Cool It, Orville Redenbacher*.

3-9 Backboard display; *Cool It, Orville Redenbacher.*

3-10 Backboard display; *Cool It, Orville Redenbacher.*

How would Vy improve her project? The first thing that occurred to her was to conduct more testing to further validate her results and conclusions. Another project? Perhaps, but if she does one, Vy is ready to try something more scientifically challenging.

What did she get out of it? The project gave her scientific experience through conducting the experiment, and artistic expression through creating the backboard. She also won her first license—poetic license—by titling the project "Cool It, Orville Redenbacher," when she actually used Jolly Time popcorn!

GREEN THUMB

Most people have plants around the house, and many of us would love to have them grow green and strong instead of brown and soggy. This next project suggests that water drainage may be the key.

Jennifer Lynne Ade was always interested in plants. When her teacher assigned a science project, she saw a presentation featuring a botany project, and realized that she could work in one of her favorite areas. At first, she decided to do an experiment testing different types of soils. However, because a classmate had already submitted that topic, her teacher helped her develop another idea, an experiment relating water drainage to plant growth.

Jennifer conducted her research over a period of five to six weeks. She used her local library and the school library, as well as several encyclopedias and plant books at home. She also got information from the backs of seed packages.

As Jennifer did research, she took notes on sheets of paper. When she had collected all her information, she identified similar facts and copied them to separate sheets, which organized the material by subject. She did not need a written outline because the note sheets by category were sufficient. Once the material was organized, Jennifer wrote the paper in about three weeks. "Because I don't like to write," said Jennifer, "the research paper was the worst part of the project."

Based on her research, Jennifer developed the question and hypothesis shown in FIG. 3-11. The variables and controls for the project are shown in FIG. 3-12, and the experimental and control groups are shown in FIG. 3-13.

Jennifer's experiment was conducted entirely at home, using the materials listed in FIG. 3-14. She grew radishes because they can flourish at any time of year in Southern California, and also because they sprout rapidly, within one week after planting.

Question	Hypothesis
Does water drainage adversely affect plant growth?	A moderate rate of drainage promotes plant growth.

3-11   Question and hypothesis; *Effects of Drainage on Plants.*

Variables	Controls

Experimental
          Drainage
 	
Measured
          Growth of plants
	• Amount of water
	
	• Amount of sunlight
	
	• Type and amount of soil

3-12   Variables and controls; *Effects of Drainage on Plants.*

Experimental Groups	ontrol Group
1. Plants with 2 drainage holes 2. Plants with 4 drainage holes 3. Plants with 6 drainage holes	None

3-13   Experimental and control groups; *Effects of Drainage on Plants.*

The entire experiment ran for four weeks during November and December, 1989. During the first week, however, Jennifer prepared the experiment. She divided the cups into three groups, labeled *fast*, *medium*, and *slow*, and numbered every cup in each group. In the bottom of the slow cups, she punched 2 holes, in the bottom of the medium cups, she punched 4 holes, and in the bottom of the fast cups, she punched 6 holes. Into each cup, she placed 1 cup of soil, two seeds, and then an additional 1/2 cup of soil. She placed all seedlings in one small area in the backyard, to ensure that each plant would receive the same amount of sunlight and rain.

Once each week during the remaining three weeks of the

Materials

1. 120 styrofoam cups

2. 240 Cherry Belle radish seeds

3. 1-cup measuring cup

4. Nail set to punch drainage holes in cups

5. Potting soil

6. 1-foot ruler

3-14 List of materials; *Effects of Drainage on Plants.*

experiment, Jennifer measured and recorded the height of each plant, and watered each plant with $1/2$ cup water. The procedures are summarized in FIG. 3-15.

At the end of the experimental period, Jennifer used her logs to build tables showing the growth of each plant group. Figure 3-16

Procedures

1. Separated cups into three groups

2. Labelled each group: FAST, MEDIUM, SLOW, and numbered each cup in the group (1-40)

3. Punched holes in bottoms of cups

   - 6 in FAST
   - 4 in MEDIUM
   - 2 in SLOW

4. Into each cup, placed:

   - 1 cup soil
   - 2 radish seeds
   - 1 cup soil

5. Each week:

   - Measured and recorded height of each plant.
   - Watered each plant with 1/2 cups of water.

3-15 Procedures; *Effects of Drainage on Plants.*

shows the growth of the plants with slow drainage, FIG. 3-17 shows the growth of the plants with medium drainage, and FIG. 3-18 shows the growth of the plants with fast drainage. She also calculated the average growth for each group and created a combined graph, (FIG. 3-19).

PLANT GROWTH - SLOW WATER DRAINAGE

Height in Inches

PLANT	WEEK 1	WEEK 2	WEEK 3	PLANT	WEEK 1	WEEK 2	WEEK 3
1	0	0	0	21	0	0	0
2	0	0	0	22	0	0	0
3	0	0	0	23	0	0	0
4	0	0	0	24	0	0	0
5	0	0	0	25	0	0	0
6	0	0	0	26	0	0	0
7	0	0	0	27	0	0	0
8	0	0	0	28	0	0	0
9	0	0	0	29	0	0	0
10	0	0	0	30	0	0	0
11	0	0	0	31	0	0	0
12	0	0	0	32	0	0	0
13	0	0	0	33	0	0	0
14	0	0	0	34	0	0	0
15	0	0	0	35	0	0	0
16	0	0	0	36	0	0	0
17	0	0	0	37	0	0	0
18	0	0	0	38	0	0	0
19	0	0	0	39	0	0	0
20	0	0	0	40	0	0	0

3-16 Growth with slow drainage.

The results showed that the plants with the slowest drainage produced no growth at all, and the plants with the fastest water drainage had the greatest plant growth. Jennifer concluded that with slower drainage, more water remained in the soil and drowned the plants.

Jennifer used a pegboard for her display, which she covered with green fabric. For titles, she stencilled lettering, cut out the strips, and pasted them on the backboard. Besides the written material, Jennifer added a seed packet on the display. The backboard is displayed in FIG. 3-20.

On the whole, although she did not like writing the research

PLANT GROWTH - MEDIUM WATER DRAINAGE

Height in Inches

PLANT	WEEK 1	WEEK 2	WEEK 3	PLANT	WEEK 1	WEEK 2	WEEK 3
1	0	0	0	21	0	1/2	1
2	0	0	0	22	0	0	0
3	0	1/4	1/2	23	0	1/2	1
4	1/4	1/2	1	24	0	0	0
5	1/4	3/4	1	25	0	0	1/4
6	1/4	1/2	1	26	0	0	0
7	3/4	1 3/4	2	27	0	3/4	1
8	1/4	3/4	1	28	0	1/2	3/4
9	1/2	1	1 1/2	29	0	1/2	1
10	0	1/2	1	30	0	0	1/4
11	0	3/4	1	31	0	0	0
12	0	1/4	1/2	32	0	0	0
13	1/4	1/2	3/4	33	0	0	0
14	0	3/4	1	34	1/4	1/4	1/2
15	0	0	1/4	35	0	0	0
16	0	1/2	1	36	0	0	0
17	1/4	3/4	1	37	0	0	0
18	0	1/4	1/2	38	0	0	0
19	0	3/4	1 1/2	39	0	0	0
20	0	3/4	1	40	0	0	0

3-17 Growth with medium drainage.

paper, she enjoyed the project. "The best part," she said, "was taking care of the plants." However, Jennifer felt that if she had started earlier, she could have conducted the experiment over a longer period of time, and perhaps had more conclusive results.

If she does another project, she will pick a different topic. "I've already done this," she said, "but the next one will also be about plants."

RUSTY NAILS

Another project which uses some common household items is LeMar Slater's chemistry experiment, entitled "Will Certain Household Items Prevent Rust." LeMar's project tested four common household products for their sealant, rust preventative properties.

To execute his experiment, LeMar coated ordinary nails with four substances—nail polish, glue, motor oil, and cooking oil—leaving a fifth group uncoated, as the control group. He weighed each of the five groups of nails, soaked them in water for five days,

PLANT GROWTH - FAST WATER DRAINAGE

Height in Inches

PLANT	WEEK 1	WEEK 2	WEEK 3	PLANT	WEEK 1	WEEK 2	WEEK 3 PLANT
1	0	0	0	21	0	1/4	1/2
2	0	0	0	22	0	0	1/4
3	0	0	0	23	1/2	1	1 1/2
4	0	0	1/4	24	1/4	3/4	1
5	1/2	3/4	1	25	1	2	2 1/4
6	0	0	1/4	26	1/4	1/2	1
7	0	0	0	27	1/2	1	1 1/4
8	0	0	0	28	1/4	3/4	1
9	0	0	0	29	1/4	3/4	1
10	0	0	0	30	1/2	1	1 1/4
11	1/2	3/4	1	31	0	1/2	1
12	1/4	3/4	1	32	1/4	3/4	1
13	0	1/4	1/2	33	1/4	3/4	1
14	0	0	0	34	1 1/4	2 1/2	3
15	1/4	1/2	3/4	35	1	1 3/4	2
16	0	0	0	36	1/4	1	1 1/2
17	1/4	3/4	1	37	0	1/2	1
18	0	0	0	38	1/4	3/4	1
19	0	0	0	39	1/2	1	1 1/2
20	0	0	1/4	40	3/4	1 /12	2

3-18 Growth with fast drainage.

weighed them again, and noted the results. The nails which were lightest in weight, the group coated in nail polish, was the group that had the least rust. He therefore concluded that nail polish was the best sealant. LeMar's backboard, shown in FIGS. 3-21 through 3-23, showed not only the question, hypothesis, procedures, and other written materials, but had the rusted nails mounted on the panel.

HELPFUL HINTS

☐ Remember the KISS principle (Keep It Simple, Stupid). Sometimes, the simplest ideas are best.

☐ Do your experiment a sufficient number of times (or for a long enough period of time) to get adequate results.

☐ Including part of your experiment on your backboard can make your display more interesting.

AVERAGE PLANT GROWTH

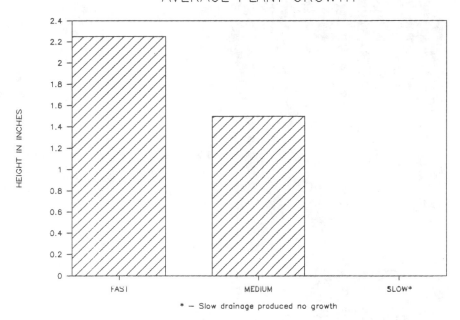

* — Slow drainage produced no growth

3-19 Combined graph.

3-20 Backboard; *Effects of Drainage on Plants*.

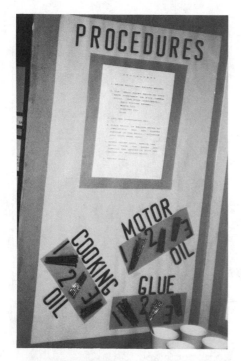

3-21   Backboard; *Preventing Rust.*

3-22   Backboard; *Preventing Rust.*

3-23 Backboard; *Preventing Rust*.

Chapter 4

Computer science

*C*omputing is an area all its own. If you're hung up on hardware, passionate about programming, or sensitive to software, a computer science project might be just what you're looking for. A computer science project is quite different from those in botany, chemistry, or medical science. Instead of conducting an experiment, a computer science project generally develops a new product, such as a programming language or a new type of software. Therefore, the entire method of doing a computer science project, from the research to the procedures, follows a different path.

DO YOU SPEAK BASIC?

Davis Houlton had no question in his mind that he'd do a computer science project. When the semester began, he was working on a spelling program and thought he'd do some further work on that. However, a friend of his, whose native language is Spanish, was having a problem with his computer classes, because programming requires learning a new language in addition to English. Since computer literacy is important to success in all fields, Davis reasoned that a method of helping non-English speakers to learn programming would be helpful. Besides Spanish—which is a common second language for students in San Diego, located on the Mexican border—Davis also chose to translate Tagalog and Vietnamese, which are native languages for many other students at his school. He decided not to translate French or German because the ability to program using those languages already exists.

The project plan was to translate Spanish, Tagalog, and Vietnamese into BASIC, the programming language that is taught at his school. The project question and hypothesis are shown in FIG. 4-1.

Question	Hypothesis
Can I design and code a system that allows a user to enter a program in his or her native language?	I can use variations of the READ/WRITE statements in BASIC to develop a system that allows a user to enter a program in his or her native language.

4-1   Question and hypothesis; *XLATE.*

To do his background research, Davis first consulted several local branch libraries, but found almost nothing available that related to his topic. He checked the card catalog, and found that even the central library had very few current resources. In computer science, which is a rapidly changing field, reference materials become out-of-date very quickly, and new sources of information are published just as fast. However, Point Loma Nazarene College, which was close to his home, had a great deal of material on the BASIC programming language, including how it got started and how it works. (Davis's bibliography appears in the back of this book.)

While doing his research, Davis entered each fact as well as its source and category, using Microsoft Windows. When he had accumulated all his information, he used the note-card feature of Windows to sort the facts according to category, format them as index cards, and print the cards. (In Davis's science class, note cards were required as a project milestone.) Then, he used the cards as the basis of the background research paper, which he wrote using WordPerfect 5.0.

When developing a complex computer project, it is important to start by designing the system and outlining the steps involved. To develop the system design, Davis created a block diagram, shown in FIG. 4-2. He also produced a pseudocode design, which is an English language outline of a program or system, shown in FIGS. 4-3, 4-4, 4-5, and 4-6.

The system, named XLATE, consists of three programs. The startup program allows a user to begin using the XLATE system. Through a menu, shown in FIG. 4-7, the user tells the system the language he or she wants to enter, and whether he or she will use a

XLATE PROGRAM

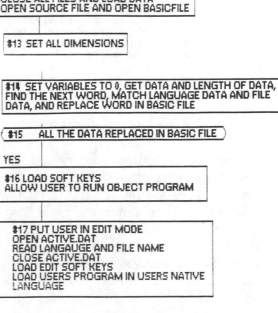

YES   #1 SET UP PARAMETERS   NO
- USER DECIDES ON LANGUAGE
- IS THERE AN ACTIVE.DAT?

#3 DOES LANGAUGE IN   NO
ACTIVE.DAT=LANGAUGE
USER SELECTED?

YES

#4 ACTIVE.DAT VERIFICATION   #2 ACTIVE.DAT CREATION
IS FILENAME CORRECT?   NO   -LANGAUGE   -FILENAME

YES

NO   #5 EDIT RUN MENU EDIT
USERS PROGRAM?

YES

#6 LOAD SOFT KEYS
USER TYPES IN PROGRAM

#7 OPEN ACTIVE.DAT
READ LANGUAGE AND FILENAME
CLOSE ACTIVE.DAT

#8 IS THERE AN OLD FILE?   YES   #9 DELETE THE
OLD FILE

NO

#10 IS THERE A SOURCE FILE?   YES   #11 RENAME THE SOURCE
FILE AS THE OLDFILE

NO

#12 RENAME T.SRC AS THE SOURCEFILE
CLOSE ALL FILES AND LOAD DATA
OPEN SOURCE FILE AND OPEN BASICFILE

#13 SET ALL DIMENSIONS

#14 SET VARIABLES TO 0, GET DATA AND LENGTH OF DATA,
FIND THE NEXT WORD, MATCH LANGUAGE DATA AND FILE
DATA, AND REPLACE WORD IN BASIC FILE

#15   ALL THE DATA REPLACED IN BASIC FILE

YES

#16 LOAD SOFT KEYS
ALLOW USER TO RUN OBJECT PROGRAM

#17 PUT USER IN EDIT MODE
OPEN ACTIVE.DAT
READ LANGAUGE AND FILE NAME
CLOSE ACTIVE.DAT
LOAD EDIT SOFT KEYS
LOAD USERS PROGRAM IN USERS NATIVE
LANGUAGE

4-2 Block diagram of XLATE program.

NOTES

==

PROGRAM INFORMATION:

The three runtime programs are:

MENU.BAS = This is the start of the program. It asks the
user for speaking language and filename. From
here the user has the capability to go to the
XLATE program after the user types in his
program.

XLATE.BAS= This is main program. This program translates
the users program from foreign to english so
BASIC can understand the new program. From
here the user can go to the EDIT program.

EDIT.BAS = The soft keys are loaded and the users'
program is loaded in the users' native
language. From here the user can go to
XLATE.

There are three data programs. They are:

SPANISH.DAT = This program contains all the data needed
for translation from spanish to english.

TAGALOG.DAT = This program contains all the data needed
for translation from Tagalog to English.

VIET.DAT   = This program contains all the data needed
for translation from Vietnamese to English.

There are four main user programs. They are:

ACTIVE.DAT = This contains the active file the user is
using and the current language.

T.SRC   = This is the temporary program to be used by
the user. After the user is finished with
MENU.BAS and types up his program, the user
presses the key F3 and the his/her program
is saved as T.SRC.

*.OLD   = This file is the old version of the active
file. This is used in file testing.

*.BAS   = This is what the final edition of the users
program is going to look like.

4-3 Pseudocode for XLATE.

new or old file. Part of the startup program, called MENU.BAS, is
shown in FIG. 4-8.

Once the program, or source, was entered into the computer,
the parser translation routine found each word by searching for a
series of characters between two blanks. "That was the hardest
part of the system to program correctly," said Davis.

Once the word was found, the program checked it against a
master list for the designated language to find the word's translation. Portions of the master lists are shown in Spanish in FIG. 4-9,
Tagalog in FIG. 4-10, and Vietnamese FIG. 4-11. If a word was not
found on the master list, it remained in its original language. To

PROGRAM DESCRIPTION:

Here is the PDL of XLATE:

The first part of XLATE.BAS does the following:
- Opens "ACTIVE.DAT" for input as field #1
- Reads the data in "ACTIVE.DAT"
- Closes "ACTIVE.DAT"

The second part of XLATE.BAS does the following:
- Checks for an error, and if there is one, go to the ERROR handling routine
- Checks to see if there is an OLD FILE. If there is, the OLD FILE is deleted from the disk.
- Checks for another error. If there is one, the program goes to the SECOND ERROR handling routine
- Opens the source file, then changes the name of the source file to OLD FILE
- The main file "T.SRC" is named after the source file
- Execute the DATA loading routine
- All files are closed, then the source file is opened in field #1.
- The BASIC FILE is opened in field #2
- The dimensions of all subscript variables are assigned
- Data is read into memory. If there is no more data, then execute END

The third part of my program is does the following:
- Sets all variables to correct setting
- Gets the length of data
- Grabs a PIECE of the data
- Check to see if the PIECE is a variable or a reserved word
- Put the data into the BASIC FILE in field number 2
- Get more data. If there is no more data, then execute END
--- ROUTINES ---
ERROR =
- Return to where the SOURCE FILE is being opened

SECOND ERROR =
- Return to where the SOURCE FILE is being named

DATA =
- Load the correct data set according to language
- Return to point left off
END =
- Set up the special keys
- Print out what the special keys are
- Load the BASIC FILE
- End of XLATE

4-4 Pseudocode for XLATE.

develop the master list, Davis consulted computer programmers who speak Spanish and Vietnamese, and a teacher who speaks Tagalog. For further reference, he also used dictionaries in all three languages. Portions of the XLATE.BAS program are shown in FIGS. 4-12 and 4-13.

The edit program loaded the *soft keys*, listed in FIG. 4-14. Soft keys allow the programmer to use function keys to enter some common commands, instead of typing in the entire command. The program then loaded and ran the BASIC program, locating and

```
        Here is the PDL of MENU.BAS:

Here is the first part of MENU.BAS:
        - Wait for user to type in the correct number for
speaking language.
        - If an error happens in the next part, execute the SPEECH
routine.
        - Opens and reads ACTIVE.DAT
        - If the language the user entered and the language in
ACTIVE.DAT are different, then execute the SPEECH routine.
        - If the language is SPANISH then goto SPANISH FILE routine.
        - If the language is TAGALOG then goto TAGALOG FILE routine.
        - If the language is VIETNAMESE then goto VIETNAMESE FILE
        routine.

ROUTINES -
SPANISH FILE -
        - Print out file screen
        - Verify that the specs are proper
        - If the file name is incorrect then execute SPANISH
ACTIVE.DAT routine.
        - If the file name is correct then execute SPANISH EDIT

SPANISH ACTIVE.DAT -
        - Print out active.dat file screen
        - Wait until filename is typed and correct
        - Write language and filename to ACTIVE.DAT
        - Continue to SPANISH EDIT

SPANISH EDIT -
        - Print out edit screen
        - Wait for numbered input
        - If 1 is entered then run XLATE
        - if 2 is entered then goto the SPANISH EDIT KEYS routine.

SPANISH EDIT KEYS -
        - Set soft keys
        - Get user ready to program
        - End

TAGALOG FILE -
        - Print out file screen
        - Verify that the specs are proper
        - If the file name is incorrect then execute TAGALOG
ACTIVE.DAT routine.
        - If the file name is correct then execute SPANISH EDIT

TAGALOG ACTIVE.DAT -
        - Print out active.dat file screen
        - Wait until filename is typed and correct
        - Write language and filename to ACTIVE.DAT
```

4-5 Pseudocode for XLATE.

displaying any errors that it found. Portions of EDTSRC.BAS are shown in FIG. 4-15.

To assist users, Davis also developed instructions and menu screens in the three languages. Samples of these menus, translated into Spanish, are illustrated. Figure 4-16 shows the ready menu, FIG. 4-17 shows the screen that appears when the program is loaded, and FIG. 4-18 illustrates the user's choice of whether to edit or run a program. Figure 4-19 shows the screen that asks the user to enter a file name, and FIG. 4-20 lets the user verify the file name.

4-5 Continued.

```
            - Continue to TAGALOG EDIT

TAGALOG EDIT -
      - Print out edit screen
      - Wait for numbered input
      - If 1 is entered then run XLATE
      - if 2 is entered then goto the TAGALOG EDIT KEYS routine.

TAGALOG EDIT KEYS -
      - Set soft keys
      - Get user ready to program
      - End

VIETNAMESE FILE -
      - Print out file screen
      - Verify that the specs are proper
      - If the file name is incorrect then execute VIETNAMESE
ACTIVE.DAT routine.
      - If the file name is correct then execute SPANISH EDIT

VIETNAMESE ACTIVE.DAT -
      - Print out active.dat file screen
      - Wait until filename is typed and correct
      - Write language and filename to ACTIVE.DAT
      - Continue to VIETNAMESE EDIT

VIETNAMESE EDIT -
      - Print out edit screen
      - Wait for numbered input
      - If 1 is entered then run XLATE
      - if 2 is entered then goto the VIETNAMESE EDIT KEYS routine.

VIETNAMESE EDIT KEYS -
      - Set soft keys
      - Get user ready to program
      - End

SPEECH -
      - If the language is spanish then goto SPANISH ACTIVE.DAT
      - If the language is Tagalog then goto TAGALOG ACTIVE.DAT
      - If the language is Vietnamese then goto VIETNAMESE
ACTIVE.DAT
```

4-5 Pseudocode for XLATE.

```
      This is the EDIT PDL:

This is what the first part of EDIT.BAS does:
      - Opens and reads ACTIVE.DAT
      - If the language is  Spanish, then execute SPANISH KEYS
      - If the language is Tagalog, then execute TAGALOG KEYS
      - If the language is Vietnamese, then execute VIETNAMESE KEYS

ROUTINES==
SPANISH KEYS-
      - Load soft keys
      - End

TAGALOG KEYS-
      - Load soft keys
      - End

VIETNAMESE KEYS-
      - Load soft keys
      - End
```

4-6 Pseudocode for XLATE.

```
1-Espanol
2-Tagalog
3-Vietnamese
```

4-7 Startup menu.

```
1    REM   MENU.BAS
2    REM   THIS PROGRAM REQUESTS THE LANGUAGE AND THE LANGUAGE SOURCE FILE
3    REM   INPUT:
4          USER PROMPT: LANGUAGE AND FILENAME
5          ACTIVE.DAT: THE PREVIOUS SESSION LANGUAGE AND FILENAME
6    REM   OUTPUT:
7          ACTIVE.DAT: THE CURRENT LANGUAGE AND FILENAME
8          T.SRC: THE SOURCE LANGUAGE PROGRAM TO BE TRANSLATED
9    :
10   CLS
20   WIDTH 80
30   COLOR 9,4,9:CLS
40   KEY OFF
44   :
45   REM #1 - MAIN MENU
50   PRINT:LOCATE 2,36:PRINT"XLATE SPANISH/VIETNAMESE/TAGALOG
60   PRINT:PRINT"=====================================================
================"
64   :
65   REM #1 - 1 = SPANISH SELECTION. 2 = TAGALOG SELECTION 3 = VIETNAMESE
SELECTION
70   PRINT:PRINT"1-Espanol
80   PRINT:PRINT"2-Tagalog
90   PRINT:PRINT"3-Vietnamese
100  A$=INKEY$:IF A$="" THEN GOTO 100
104  :
105  REM #1 - SET SPEECH CODE
110  IF A$=1 THEN LANG$+"SPANISH":SPEECH=1
120  IF A$=2 THEN LANG$+"TAGALOG":SPEECH=2
130  IF A$=3 THEN LANG$+"VIETNAMESE":SPEECH=3
140  IF A$<>"1" AND A$<>"2" AND A$<>"3" THEN GOTO 100
```

4-8 Portion of MENU.BAS.

While developing the three programs, Davis used the procedures outlined in FIG. 4-21. However, in his science project notebook, the list of procedures is considerably longer and more detailed.

```
7990 REM SPANISH.DATA
7991 REM This file is the Spanish to BASIC reserved word translation
8000 RESTORE:DIM E$(169):DIM F$(169):FOR I TO (169): READ F$(I): READ E$(I):
NEXT I: GOTO 140:
9000 DATA "ABS","ABS"
9010 DATA "YA","AND"
9020 DATA "ASC","ASC"
9030 DATA "ATN","ATN"
9040 DATA "AUTOMATICO","AUTO"
9050 DATA "TONO","BEEP"
9060 DATA "BLOAD","BLOAD"
9070 DATA "BSAVE","BSAVE"
9080 DATA "LLAMA","CALL"
9090 DATA "CDBL","CDBL"
9100 DATA "CADENA","CHAIN"
9110 DATA "CAMDIR","CHDIRL"
9120 DATA "LET$","CHR$"
9130 DATA "CINT","CINT"
9140 DATA "CIRCULO","CIRCLE"
9150 DATA "ACLARA","CLEAR"
9160 DATA "CIERRA","CLOSE"
9170 DATA "BORRAM","CLS"
9180 DATA "COLOR","COLOR"
9190 DATA "COMUN","COMMON"
```

4-9 Portion of Spanish master list.

```
7990 REM TAGALOG.DATA
7991 REM This file is the Tagalog to BASIC reserved word translation
8000 RESTORE:DIM E$(169):DIM F$(169):FOR I TO (169): READ F$(I): READ E$(I):
NEXT I: GOTO 140:
9000 DATA "TAWAG","CALL"
9010 DATA "BILOG,"CIRCLE"
9020 DATA "MALINAW","CLEAR"
9030 DATA "SARADO","CLOSE"
9040 DATA "KULAY","COLOR"
9050 DATA "PUNGKARANIWAN","COMMON"
9060 DATA "PETCHA","DATE$"
9070 DATA "ASLIN","DELETE"
9080 DATA "AT","AND"
9090 DATA "BURAHIN","ERASE"
9100 DATA "BUKIOL","FIELD"
9110 DATA "KUNG","IF"
9120 DATA "PATAYIN","KILL"
9130 DATA "NATILIRA","LEFT$"
9140 DATA "GUHIT","LINE"
9150 DATA "TINGNANOHAMAPIN","LOCATE"
9160 DATA "MAGIISAOMAGSASAME","MERGE"
9170 DATA "PANGALAN","NAME"
9180 DATA "BAGO","NEW
9190 DATA "SUNSUNOD","NEXT"
```

4-10 Portion of Tagalog master list.

```
7990 REM VIET.DATA
7991 REM This file is the Vietnamese to BASIC reserved word translation
8000 RESTORE:DIM E$(169):DIM F$(169):FOR I TO (169): READ F$(I): READ E$(I):
NEXT I: GOTO 140:
9000 DATA "TUUETDOI","ABS"
9010 DATA "VA","AND"
9020 DATA "ASC","ASC"
9030 DATA "ATN","ATN"
9040 DATA "AUTO","AUTO"
9050 DATA "BEEP","BEEP"
9060 DATA "BLOAD","BLOAD"
9070 DATA "BSAVE","BSAVE"
9080 DATA "GOI","CALL"
9090 DATA "CDBL","CDBL"
9100 DATA "CHAIN","CHAIN"
9110 DATA "CHDIR","CHDIRL"
9120 DATA "CHU$","CHR$"
9130 DATA "CINT","CINT"
9140 DATA "VONGTRON","CIRCLE"
9150 DATA "XOA","CLEAR"
9160 DATA "DONGLAI","CLOSE"
9170 DATA "XOAMAN","CLS"
9180 DATA "MAU","COLOR"
9190 DATA "COMMON","COMMON"
```

4-11 Portion of Vietnamese master list.

To test the XLATE system, Davis checked the master lists in Spanish, Tagalog, and Vietnamese both manually and using DOS (Disk Operating System) commands. He also developed a demonstration program, which he translated into all three languages and successfully ran.

Overall, he found that he was able to create an effective language translator in all three languages. However, the Tagalog version was somewhat less effective because the language does not contain many technical terms, and several words had to remain untranslated.

"XLATE will be most effective in the classroom," Davis said, "where there's a teacher to help the students with the actual BASIC programming."

The backboard, which featured the block diagram and portions of the actual BASIC code, as well as photos of students using the program, appear in FIG. 4-22.

If Davis decides to do another project, he has already identified several enhancements for XLATE. The most time-consuming part of the system's execution is searching the master list for the translation of each word. Currently, the program begins at the first word in the list for the selected language. XLATE may need to check the entire list of 189 items before it finds a match, or determines that no match exists. However, in any future version, Davis will use an indexing procedure. This will speed up the search, because not every word will need to be checked each time.

Davis said that one weakness, which he was unable to correct in the amount of time available, is the lack of documentation for BASIC, DOS, or the PC computer itself in languages other than English. A general system improvement would be to translate some DOS commands, as well as portions of the BASIC manual, to make it an even more effective tool.

But will he continue? That depends. "It was harder than I thought it would be," said Davis, "and took more time than I anticipated." Next year? Wait and see.

```
1    REM   XLATE.BAS
2    REM   THIS PROGRAM TRANSLATES THE FOREIGN LANGUAGE BASIC PROGRAM
3    REM   GW-BASIC AND ALLOWS THE USER TO RUN AND DEBUG THE PROGRAM.
4    REM   INPUT:
5          ACTIVE.DAT: LANGUAGE AND FILENAME
6          T.SRC: THE SOURCE LANGUAGE PROGRAM TO BE TRANSLATED
7          USER PROMPT: REQUEST TO EDIT SOURCE LANGUAGE PROGRAM
8    REM   OUTPUT:
9          FILENAME.OLD: PREVIOUS VERSION OF SOURCE LANGUAGE PROGRAM
10         FILENAME.SRC: CURRENT VERSION OF SOURCE LANGUAGE PROGRAM
11         FILENAME.BAS: TRANSLATED PROGRAM IN GW-BASIC
12   :
13   REM #7 - OPEN ACTIVE.DAT, READ LANGUAGE (LANGS) AND THE FILENAME
(FILES) FROM ACTIVE.DAT AND THEN CLOSE ACTIVE.DAT
14   OPEN,"i",#1,"ACTIVE.DAT"
15   INPUT #1,"LANG$,FILE$
16   IF LANG$="SPANISH" THEN NOD=167
17   REM #8 - CHECK FOR OLDFILES (FILE$.OLD). IF OLDFILE IS NOT LOCATED
CONTINUE TO 5500
18   ON ERROR GOTO 5500
19   IF LANG$="TAGALOG" THEN NOD=73
20   IF LANG$="VIETNAMESE" THEN NOD=169
25   EXE$=".OLD":OFILES=FILES+EXE$
30   CLOSE #3
35   REM #9 - DELETE OLDFILE
40   KILL OFILES
44   :
45   REM #10 - CHECK FOR SOURCE FILE (FILE$.SRC). IF SOURCE FILE IS NOT
LOCATED, CONTINUE TO 5520
50   ON ERROR, GO TO 5500
```

4-12 Portion of XLATE.BAS.

```
1565 REM  #14 - MATCH DATA FROM SOURCE FILE AGAINST LANGUAGE DATA
1580 IF BS=1 AND TYPE(BS)=1 THEN A$=MID$(B$,START(BS)NUMBER(BS)) ELSE
GOTO 1640
1600 FOR I=1 TO NOD:IF A$=F$(I) THEN A$=E$(I):GOTO 1740
1601 NEXT I
1602 GOTO 1740
1640 IF BS=1 AND TYPE(BS)=0 THEN A$=MID$(B$,START(BS),NUMBER(BS)):GOTO 1740
1660 IF BS<>1 AND TYPE(BS)=1 THEN WORD$=MID$(B$,START(BS),NUMBER(BS))
ELSE GOTO 1720
1680 FOR I=1 TO NOD:IF WORD$=F$(I) THEN WORD$=E$(I):A$=A$+WORD$:GOTO 1740
1681 NEXT I A$=A$+WORD$:GOTO 1740
1720 IF BS<>1 AND TYPE(BS)=) THEN WORD$=MID$(B$,START(BS),NUMBER(BS))
1740 NEXT BS
1744 :
1760 PRINT#2,A$
1780 GOTO 170
4999 REM #12 - LOAD LANGUAGE DATA AND RETURN TO 140
5010 IF LANG$="SPANISH" THEN CHAIN MERGE "TAGALOG.DAT",8000,ALL
5010 IF LANG$="TAGALOG" THEN CHAIN MERGE "VIET.DAT",8000,ALL
5020 IF LANG$="VIETNAMESE" THEN CHAIN MERGE "VIET.DAT",8000,ALL
5040 GOTO 140
```

4-13 Portion of XLATE.BAS.

Soft Keys
The list of editing soft keys are as follows:

F1 = LIST	Lists the program
F2 = RUN XLATE	Runs XLATE
F3 = AUTO	Sets automatic line numbering on
F4 = SAVE T.SRC	Saves the user's program as T.SRC
F5 = LLIST	Prints out the program on the printer
F6 = RENUM	Renumber lines
F7 = DELETE	Delete lines
F8 = EDIT	Edit a line

The runtime soft keys are as follows:

F1 = PRINT	Print out something on the screen
F2 = RUN	Run the program
F3 = SYSTEM	Go to DOS
F4 = RUN EDIT	Run the EDIT.BAS program
F5 = CONT	Continue a STOPped program
F6 = FILES	Lists all the files
F7 = TRON	Trace on
F8 = TROFF	Trace off

4-14 Soft keys for XLATE.

```
1     REM   EDTSRC.BAS
2     REM   THIS PROGRAM PROCESSES THE USER SELECTION OF "SRC" FROM THE
3     REM   TRANSLATE PROGRAM.  THE USER CAN EDIT THE SOURCE LANGUAGE
4     REM   PROGRAM.
5     REM   INPUT:
6           ACTIVE.DAT: LANGUAGE AND FILENAME
7           FILENAME.SRC: THE SOURCE LANGUAGE PROGRAM
8           USER PROMPT: REQUEST TO TRANSLATE & RUN SOURCE PROGRAM
9     :
10    CLOSE
11    SCREEN 2 0
15    REM #17 - OPEN ACTIVE.DAT, READ LANGUAGE FILE & FILE NAME FROM
ACTIVE.DAT AND CLOSE ACTIVE.DAT
20    OPEN"I", #1,"ACTIVE.DAT"
40    INPUT#1,"LANG$,FILE$
40    CLOSE
44    :
45    REM #17 - SET USERS FILE NAME
46    EXE$=".SRC":FILE$+EXE$
50    IF LANG$="SPANISH" THEN GOTO 5000
60    IF LANG$="TAGALOG" THEN GOTO 2000
70    IF LANG$="VIETNAMESE" THEN GOTO 3000
1999  :
2000  CLS
2001  REM#17 - LOAD SOFT KEYS
2010  KEY OFF
2020  KEY 3,"AUTO"+CHR$(13)
2030  KEY 1,"LIST"+CHR$(13)
2040  KEY 5,"LLIST"
2050  KEY 6,"RENUM"
2060  KEY 4,"SAVE"+"T.SRC"+CHR$(34)+",A"+CHR$(13)
2065  KEY 2,"RUN"+"XLATE.BAS"+CHR$(13)
2070  KEY 7,"DELETE"
2075  KEY 8,"EDIT"
2080  LOCATE 25,1:PRINT"1-LIST  2-TAKBO  3-AUTO  4-ITABI  5-LLIST  6-RENUM
7-ALISIN  8-EDIT"
2100  PRINT"TAGALOG":PRINT:PRINT:PRINT:PRINT:PRINT:PRINT:PRINT
2105  REM#17 - LOAD FILE.SRC
2110  LOAD FILE$
2999  :
```

4-15 Portion of EDTSRC.BAS.

```
Listo a Programar




Ok

1-IMPRIME  2-CORRE  3-SYSTEMA  4-SRC  5-CONT  6-ARCHIVOS  7-TRON  8-TROFF
```

4-16 Spanish menu: ready.

```
DAVIS.BAS LOADED->
Ok

1-IMPRIME  2-CORRE  3-SYSTEMA  4-SRC  5-CONT  6-ARCHIVOS  7-TRON  8-TROFF
```

4-17   Spanish menu: Program loaded.

```
REDACTAR / CORRE

1 - REDACTAR
2 - CORRE

FILE:DAVIS.SRC
```

4-18   Spanish menu: Edit or run.

```
CREAR ACTIVE.DAT

Registrar el nombre de la ficha
```

4-19   Spanish menu: Requesting file name.

```
     NOMBRE DE LA FICHA:  DAVIS

1 - Es correcto el nombre de la ficha
2 - Es incorrecto el nombre de la ficha
```

4-20   Spanish menu: Verifying file name.

Procedures
1. Compiled list of BASIC reserved words
2. Acquired translation of BASIC reserved words in Spanish, Tagalog, and Vietnamese
3. Created flowchart for the XLATE system
4. Designed program requirements for the programs, based on the BASIC handbook
5. Designed data files (master lists) in three languages
6. Converted master lists to ASCII, then used DOS commands to check for accuracy and completeness
7. Coded three programs, MENU.BAS, XLATE.BAS, and EDTSRC.BAS
8. Coded a sample program used to test XLATE system

4-21 Procedures; *XLATE*.

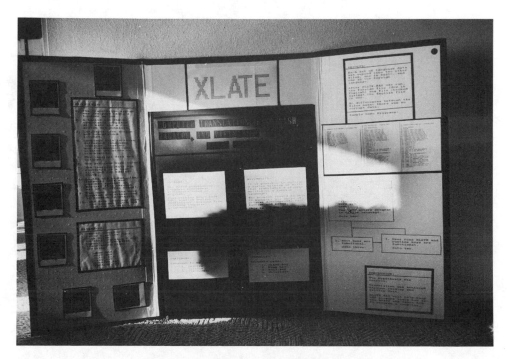

4-22 Backboard display; *XLATE* project.

HELPFUL HINTS

☐ Plan your system before you begin to program. Using block diagrams or pseudocode can help to organize your project and keep you on track.

☐ Use all the tools that are available on your computer to write your research paper, develop your statistics, or enhance your backboard.

☐ Computer science projects can be short and simple or long and complex. Before you begin try to analyze your skills and your desire to be involved with the project. Don't bite off more than you can (comfortably) chew.

☐ Davis had a few words of wisdom for anyone who makes it to the science fair. "Make sure you know your material," he said, "because if the judges are interested in your project, they'll ask questions."

Chapter **5**

All about us

Medical science is a field of big dreams—finding the cause of birth defects or the cure for a deadly disease. Most of these discoveries, however, are not the result of sudden inspiration or a quick bump on the head. Instead, they are made after years of hard work doing patient research and experimentation. While you're spending a few months doing a science project, it's very unlikely that you'll have the resources, the know-how, or the time to address those kinds of issues.

But there are important, though less spectacular, areas of medical science that you can explore to create a successful and rewarding science project. These next two projects show how, by doing a simple test on a group of classmates, you can do a project in a relatively short time.

SEEING IS BELIEVING

At first, Yolanda Lockhart was not thrilled about doing a project because science is not her primary interest. "I want to be a writer," she told me at our first meeting. However, because a project was required as part of her Life Sciences class, Yolanda set out to find a suitable topic. She couldn't seem to find anything that interested her and decided at the last minute to do an experiment on mold. Her teacher promptly discouraged her because that was an idea that had been done too many times. Back to square one.

Then, Yolanda remembered an experiment that she did in fifth grade. At that time, each student paired up with a partner to test peripheral vision. Using a board with a colored tab gradually moved from the edges towards the center, the subject looked straight ahead and noted when he or she first sighted the tab, while the partner recorded the number of degrees from the center.

Yolanda decided to expand this concept to perform the peripheral vision test on three groups of subjects—those with perfect vision, those who were nearsighted, and those who were farsighted—and compare the results.

At first, she wanted to conduct the experiment using adults and children. She soon realized however, that this would make the experiment more time-consuming and the record-keeping too complicated, so she decided she could get enough subjects by testing her own classmates.

Yolanda conducted her research using the encyclopedias in the school library. She found that, although there was a great deal of material on the eye and on vision problems in general, there was very little about peripheral vision. For this reason, she wrote her research paper concentrating on the generally available knowledge in the field.

While doing research, Yolanda simply took notes on sheets of paper. Because she had written several papers and essays before, organizing the material and writing the paper presented no problem for her. When the research paper was done, Yolanda developed the project question and hypothesis shown in FIG. 5-1. The experimental and control groups are shown in FIG. 5-2.

Question	Hypothesis
What peripheral differences exist between 20/20 vision and visually impaired?	People who wear glasses have worse peripheral vision than those who do not.

5-1 Question and hypothesis; *Peripheral Vision*.

Experimental Groups	Control Group
1. Students with 20/20 vision	None
2. Nearsighted students	
3. Farsighted students	

5-2 Experimental and control groups; *Peripheral Vision*.

Yolanda was able to do the entire experiment with simple and inexpensive materials, as shown in FIG. 5-3. To prepare for the experiment, she created the experimental field on a piece of white poster board. Using a compass and protractor, she marked off a semicircle where the subject would place his or her chin. She then measured the degrees from 0 through 90 on the left side of the board and 0 through 90 on the right side of the board, marked off in 5 degree increments. Finally, she made seven tabs of different colors out of construction paper.

Materials
1. Oak tag
2. Black marker
3. Compass
4. Protractor
5. 7 colored tabs, made from construction paper

5-3 List of materials; *Peripheral Vision.*

Yolanda conducted the experiment in her science classroom, between January 23 and February 5, 1990, as shown in her project log, FIG. 5-4. Her three subject groups were divided between male and female, and between the nearsighted, farsighted, and perfect-vision subjects.

"I had no trouble finding subjects," said Yolanda. "In fact, once my friends saw what I was doing, they volunteered to be part of my experiment." At times, a classmate helped by recording the test results while Yolanda worked with the test subject. If no one was available to assist, she was able to conduct the tests and record the data by herself.

Each subject went through the entire experiment once. The subject first placed his or her chin on the marked area, with eyes straight ahead. Yolanda placed a colored tab at the 0-degree marking on the far right side of the board, and moved along the degree markings towards 90 degrees, the center of the board. As the tab moved, the subject kept his or her eyes forward and indicated when he or she first sighted the tab, and identified the color. The test was repeated for each of the seven colored tabs for both the left and right sides of the board.

A view of the actual experiment is exhibited in FIG. 5-5. Each

DECEMBER 7, 1989- Developed an idea.

DECEMBER 17, 1989- Began research.

DECEMBER 26, 1989- Continued research.

JANUARY 13, 1990- Purchased materials to arrange peripheral vision test.

JANUARY 21, 1990- Designed a peripheral vision testing board.

JANUARY 23, 1990- Began testing.

JANUARY 24, 1990- Tested five subjects.

JANUARY 25, 1990- Tested three subjects.

JANUARY 26, 1990- Tested seven subjects.

JANUARY 29, 1990- Tested five subjects.

JANUARY 31, 1990- Tested eleven subjects.

FEBRUARY 1, 1990- Tested six students.

FEBRUARY 5, 1990- Tested five students.

FEBRUARY 8- FEBRUARY 23- Designed display board, cumulated data, and formulated research folder.

5-4 Project log.

5-5 Subject prepares for the peripheral vision experiment.

trial took approximately 15 minutes, and the findings were re-corded in her experimental log. The procedures are summarized in FIG. 5-6.

Once the experiment was complete, Yolanda averaged the scores for each student. She then constructed four graphs to dis-play the average for farsighted subjects, FIG. 5-7; nearsighted sub-

Procedures
1. Subject placed chin on marked area, with eyes forward.
2. Placed colored tab at the rightmost 0° marking on the board.
3. Gradually moved tab towards 90°, the center of the board.
4. Subject indicated when tab first sighted, and when color first identified..
5. Placed colored tab at the leftmost 0° marking on the board.
6. Gradually moved tab towards 90°, the center of the board.
7. Subject indicated when tab first sighted, and when color first identified..

5-6 Procedures; *Peripheral Vision.*

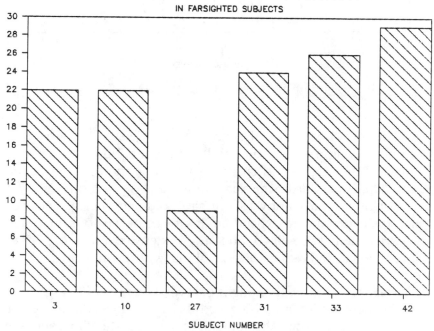

5-7 Graph of farsighted subjects.

jects, FIG. 5-8; 20/20 females, FIG. 5-9; and 20/20 males, shown in
FIG. 5-10. Based on her results, Yolanda concluded that there is a
difference in peripheral vision between those 20/20 vision and
those with visual impairment.

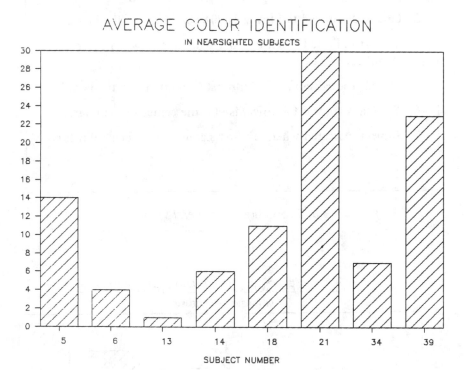

5-8 Graph of nearsighted subjects.

The backboard for the project, entitled "Comparison of
Peripheral Vision in Persons with 20/20 and Impaired Vision,"
contained the graphs and charts, as well as a photo of her experiment. The backboard panels are shown in FIGS. 5-11 through 5-13.

Yolanda already knows she'll have to do another project next
year, so she decided to expand on this idea. For the continuation of
the experiment, she will conduct more extensive research, perhaps
using medical references or consulting with professionals in the
field.

Yolanda has already thought of several ways of building on her
project. Because Yolanda observed that her subjects more readily
observed the yellow tab, she might analyze the differences in
peripheral vision based on the color of the tab. Another plan is to
use the subject groups she had originally wanted to use, adults as

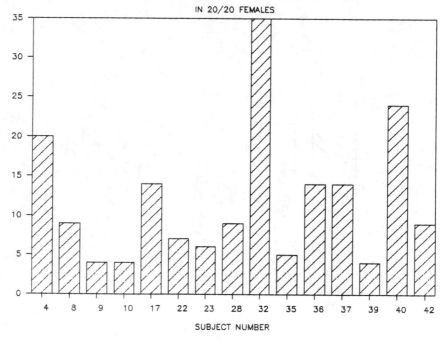

5-9 Graph of 20/20-vision female subjects.

well as children. A third possibility is to compare differences in peripheral vision between the left and right eye.

"Even though I wasn't enthusiastic to begin with," said Yolanda, "I began to enjoy the project as time went on, especially the testing. I think it will be fun to take advantage of what I've learned, and take the project a step farther."

WHERE THERE'S SMOKE . . .

When Daryl Smith learned he had to do a science project, he didn't have any particular subject in mind. In order to find a topic, he started brainstorming and listing subjects until he had a list of about ten ideas. Most of them were in the medical field, which is not surprising, given that his father is a physician.

He then talked with his science teacher and together they narrowed the list down to three ideas. Armed with his smaller list, Daryl hit the libraries to find out which topic had the most information available. Based on this preliminary research, he decided to test the effects of second-hand smoking. Until then, he hadn't read a lot about this topic, but he knew that smoking had become

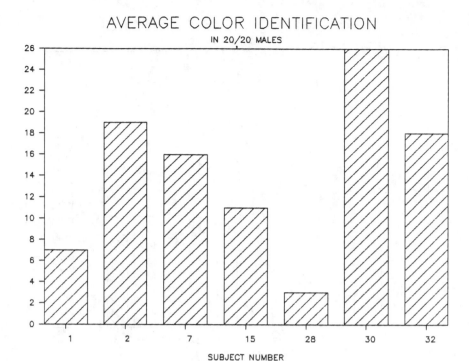

AVERAGE COLOR IDENTIFICATION
IN 20/20 MALES

5-10 Graph of 20/20-vision male subjects.

socially unacceptable. Also, some of his friends' parents smoke, and some do not, so he planned to test children of smokers and non-smokers.

Daryl did most of his research at the school library, a local branch library, and the central library. He used some references which were available at his father's office, as well as the Surgeon General's report on smoking. Daryl borrowed the books he needed, and tagged relevant pages with post-it notes. His bibliography is included in the back of this book.

He outlined his paper by writing down the various subjects he wanted to include, such as the physiology of lungs, carcinogens, and the effects of smoking. After he wrote the opening sentence for a paragraph, he went through his references to find all the facts that pertained to that particular subject, and added them to the paragraph. He wrote his background research paper on a word processing program on his computer. While conducting his research, Daryl developed the question and hypothesis shown in FIG. 5-14.

To conduct the experiment, Daryl tested 100 students—50 whose parents are smokers, and 50 whose parents are non-

5-13 Backboard; *Peripheral Vision* experiments.

5-12 Backboard; *Peripheral Vision* experiments.

5-11 Backboard; *Peripheral Vision* experiments.

Question	Hypothesis
Does involuntary smoking caused decreased lung capacity?	Children of smokers will have less lung capacity than children of non-smokers.

5-14   Question and hypothesis; *Effect of Involuntary Smoke.*

smokers. The subjects were equally divided between males and females.

Before testing, he had each subject fill out a questionnaire to make sure he or she did not have asthma or any other condition that would affect the results. To assure a constant environment, Daryl conducted all tests in the same classroom, and did no testing on excessively hot, cold, dry, or humid days. The project variables and controls are shown in FIG. 5-15, and the experimental and control groups are shown in FIG. 5-16. Daryl used the materials shown in FIG. 5-17 to conduct his experiment.

As each subject began the experiment, Daryl explained what would happen in the course of the test. He had the subject take a deep breath and exhale through the sterilized mouthpiece respiradyne, which measures lung capacity in six ways, until the subject was unable to continue. He asked the volunteer to repeat the exercise, and then he recorded the higher set of all six respiradyne measurements. A summary of experimental procedures are shown in FIG. 5-18.

When all subjects were tested, Daryl researched the meaning of all six measurements. He determined that FEV1, the forced expiratory volume in one second, was the best overall indication of lung condition and decided to base his results on that measurement.

Daryl then listed the total and mean of all the high scores (FIG. 5-19). He noted that the zero scores, shown for subjects 31 and 40 in the non-smoking group, "would mean that the people were dead." He therefore did not include these measurements when computing average or mean scores. He also graphed the scores for children of smokers and non-smokers, as shown in FIGS. 5-20 and 5-21. After analyzing the data, Daryl concluded that the children of smokers actually had higher FEV1 scores, which disproved his hypothesis.

Variables	Controls

Experimental
      Whether parents
      smoked
 	
Measured
 	
      FEV¹
	• Room temperature
	  between 72º and 80º F.
	• Subjects were free of
	  respiratory disease or
	  allergy

5-15 Variables and controls; *Effect of Involuntary Smoke.*

Experimental Groups	Control Group
Students whose parents are smokers (Must have been exposed to smoke for at least seven years)	Students whose parents are non-smokers

Note:

Both the experimental and control groups must meet the following requirements:

• Must be in Junior High School (grades 7, 8, and 9)
• Must be at rest for at least 10 minutes prior to beginning test
• Must have no recent respiratory illness.

5-16 Experimental and control groups; *Effect of Involuntary Smoke.*

One possible reason for the unexpected results was the fact that the testing was done during flu season. "Even though I eliminated anyone who was sick," said Daryl, "they might have been in an incubation period at the time."

Another factor that could have affected Daryl's experiment was that in homes where parents do smoke, Daryl did not know what their specific smoking patterns were. For example, in some homes smoking is restricted to a certain area of the house or outdoors on a patio or balcony. Also, in the homes where the parents did not smoke, he didn't know what other exposure students might have had to second-hand smoke, such as through friends and relatives who smoke.

The backboard for the project, entitled "Effect of Involuntary Smoking on Children," is displayed in FIG. 5-22.

Materials

1. One Respiradyne (Pulmonary Function Monitor), which measures lung capacity in liters, as follows:

   - FEV[1], which measures the forced expiratory volume in one second

   - FVC, the forced vital capacity

   - Peak flow

   - FEF 25-75, which measures 25-75% of the forced expiratory flow

   - Volume extra percent

   - FEV[1] / FVC

2. Four hollow mouthpieces, with flow control at the end

3. Twenty hollow mouthpiece adapters

4. One hundred alcohol swabs to sterilize the mouthpiece

5-17 List of materials; *Effect of Involuntary Smoke.*

Procedures

1. Subject seated and instructed on procedures for test.

2. Subject took mouthpiece, and after inhaling as much air as possible, exhaled through sterilized mouthpiece as forcefully as possible until unable to continue.

3. Recorded Forced Expiratory Volume in one second (FEV[1])

4. Repeated test.

5-18 Procedures; *Effect of Involuntary Smoke.*

SMOKERS		NON-SMOKERS	
1. 2.56	26. 1.52	1. 2.13	26. 1.74
2. 1.77	27. 1.38	2. 1.89	27. 1.67
3. 2.86	28. 1.95	3. 1.68	28. 2.68
4. 2.41	29. 2.94	4. 1.84	29. 1.68
5. 2.4	30. 1.18	5. 1.57	30. .81
6. 3.06	31. 1.26	6. 1.65	31. .0
7. 1.72	32. 1.44	7. 2.02	32. 3.34
8. 2.33	33. 2.02	8. 2.56	33. 2.03
9. 2.2	34. 1.24	9. 1.78	34. 2.5
10. 3.2	35. 1.52	10. 1.76	35. 1.3
11. 1.81	36. 4.24	11. 1.99	36. 1.98
12. 2.52	37. 2.2	12. 1.85	37. 1.27
13. 1.11	38. 2.42	13. 1.96	38. 1.29
14. 1.65	39. 1.86	14. 1.72	39. 2.58
15. 1.36	40. 2.82	15. 3.08	40. .0
16. 1.84	41. 1.21	16. 3.38	41. 1.99
17. 3.34	42. 2.14	17. 2.92	42. .79
18. 1.49	43. 1.82	18. 2.3	43. 2.0
19. 2.16	44. 2.78	19. .83	44. 2.74
20. 2.23	45. 1.44	20. .72	45. 1.53
21. 1.43	46. 2.96	21. 1.49	46. 2.03
22. 1.02	47. .81	22. 1.58	47. 2.0
23. 1.95	48. 1.47	23. 1.25	48. 2.45
24. 2.94	49. .88	24. 1.3	49. 1.96
25. 1.18	50. 2.3	25. 1.74	50. 2.54

TOTAL= 98.82152
MEAN = 1.9764304

TOTAL= 90.371529
MEAN = 1.8827401

All volumes were measured in liters. The higher of two scores is shown in the table above.

5-19 Mean and high scores.

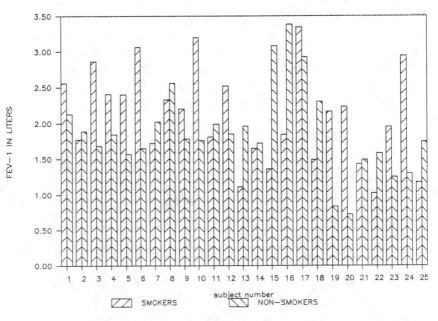

5-20 Graph of smokers and non-smokers.

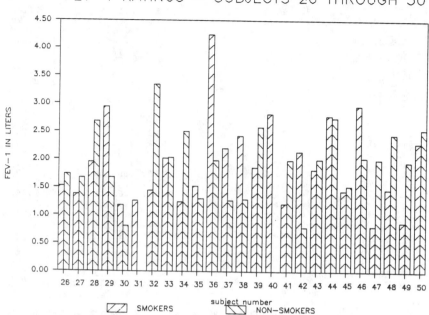

5-21 Graph of smokers and non-smokers.

Next year, Daryl intends to keep building on the same topic. He will use the same background research, but plans to use a new group of students, and also consult a specialist for the precise meaning of the measurements. He also plans to analyze all six measurements in his results and conclusions, instead of just FEV1.

Despite the results of the experiment, Daryl believes that second-hand smoking is dangerous. Next year, he hopes to prove it.

HELPFUL HINTS

☐ You may want to see how much information is available before you begin.

☐ Dig into your past. There may be some good ideas there.

☐ Look at your results! Dead subjects don't breathe!

☐ Record keeping can sometimes be tedious, but without good records there is no project.

☐ An experiment that disproves the hypothesis is as valid as one that comes out the way you expect.

☐ You can use some similar techniques for Behavioral and Social Science projects, where you need to survey a large number of people and control the results.

☐ Keeping experimental and control groups of equal size helps validate the results.

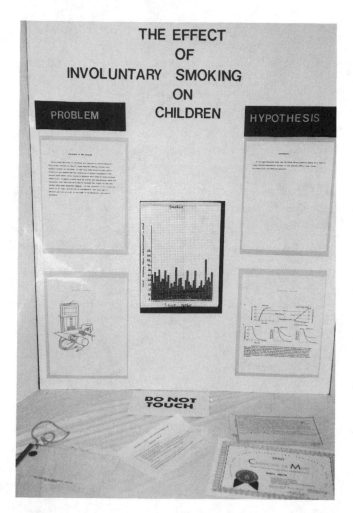

5-22 Backboard display for *Effect of Involuntary Smoke*.

Chapter **6**

Fun
and games

When you're looking for an idea for a science project, one way to start is with the things that interest you, even if they are completely unrelated to science. Are you interested in music, sports, games, or books? There might be an idea for a project buried somewhere in there.

For many of you, sports are the most important thing in your life right now. Whether you're participants or fans, sports occupy a great deal of your thoughts. Wooden bat or aluminum? The best type of cleat? Sod or Astroturf? Batting averages? Yards rushing? Any of these questions could be the start of a great science project.

HOME FIELD ADVANTAGE

You don't have to be an athlete to do a project about sports. In fact, an armchair quarterback or any other type of fan who avidly reads the sports pages, follows the home team, or quotes statistics, probably has loads of project ideas running around his or her head.

Matthew Duke is just such a sport fan and an athlete as well. As a home-team enthusiast, he wanted to find out whether the home field advantage was really fact or simply fiction. He decided on a mathematics project to determine whether the home team is a statistical favorite. The question and hypothesis are shown in FIG. 6-1.

Matthew's background paper was more of an introduction than background research. In it, he described the factors that might cause a home field advantage. In outdoor sports, the weather is often a factor because the home team is more accustomed to the local conditions. For example, the New Orleans

Question	Hypothesis
Is there a home field advantage in sports?	In any professional athletic competition, a team's performance, and likely the outcome of the game, will be affected by whether the game is played at home or away, with the home team having an advantage over the visiting team.

6-1   Question and hypothesis.

Saints, whose home-field weather would usually be quite warm, would probably be at a disadvantage playing during a snowstorm in Green Bay, Wisconsin. The playing field can be another factor that may favor the home team, especially when some fields use artificial turf and others are made of natural grass. The fatigue and jet lag involved in traveling to other cities could also cause a team to perform more poorly away from home. Finally the enthusiasm of the home team fans (or the hostility of those same fans towards the visiting team) could be another factor in the home field advantage.

By choosing basketball and soccer, sports that are played indoors on a regulation court, Matthew eliminated the weather and playing field factors from consideration.

Very few materials were used for this project, as shown in FIG. 6-2. The variables and controls, and the experimental and control groups, are shown in FIG. 6-3.

To determine how many games from each league to use for his project, Matthew sent for schedules from the NBA (National Basketball Association) and MISL (Major Indoor Soccer League). During the course of the project, which ran from November 27, 1989 through December 9, 1989, Matthew collected scores from 236 basketball games, which represented 20.77% of the NBA season, and 49 soccer games, which represented 23.56% of the MISL season. He selected a number of games as close to $1/5$ of a season as possible that would give each team an equal proportion of home and away games.

Each day, Matthew got the scores from the *San Diego Union*.

Materials
1. Daily newspaper
2. PC 386 Computer
3. dBase IV software
4. Macintosh computer
5. Microsoft Works

6-2 List of materials; *Home Field Advantage*.

Variables	Controls
Experimental Scores of MISL and NBA games **Measured** • Average home score • Average visiting score • Average winning score	Proportion of home and away games for each team.

6-3 Variables and controls; *Home Field Advantage*.

To record the data, he set up two files using a data base program, dBase IV, on his personal computer. Data base programs store and organize information so that all data or any selected portion of the data can be arranged, retrieved, and analyzed at any time. With dBase IV, Matthew created two files—one for soccer and one for basketball—that defined the information he needed to enter. For each game, he recorded the home and visiting team names and scores, and the winning score. The file structure for the soccer games is shown in FIG. 6-4, and the file structure for the basketball games is shown in FIG. 6-5.

When all the scores were entered, Matthew ran data queries on the files to list the scores and compute the average home score, visiting score, and winning score. The database query for soccer is

```
. USE SOCCER
. LIST STRUCTURE
Structure for database: C:\DBASE\SOCCER.DBF
Number of data records:      49
Date of last update    : 12/08/89
Field  Field Name   Type        Width    Dec    Index
    1  HOME_TEAM    Character      3              Y
    2  VIS_TEAM     Character      3              Y
    3  HOME_SCORE   Numeric        2              N
    4  VIS_SCORE    Numeric        2              N
    5  HOME_WIN     Character      1              Y
    6  DATE         Date           8              N
    7  WIN_SCORE    Numeric        2              Y
** Total **                      22
```

6-4   Soccer—dBase IV panel.

```
. USE BBALL
. LIST STRUCTURE
Structure for database: C:\DBASE\BBALL.DBF
Number of data records:      230
Date of last update    : 12/09/89
Field  Field Name   Type        Width    Dec    Index
    1  HOME_TEAM    Character      3              Y
    2  VIS_TEAM     Character      3              Y
    3  HOME_SCORE   Numeric        3              N
    4  VIS_SCORE    Numeric        3              N
    5  HOME_WIN     Character      1              Y
    6  DATE         Date           8              N
    7  WIN_SCORE    Numeric        3              Y
** Total **                      25
```

6-5   Basketball—dBase IV panel.

shown in FIG. 6-6, and the resulting list is shown in FIG. 6-7. The database query for basketball is shown in FIG. 6-8, and FIG. 6-9 shows a portion of the basketball season listing.

He also wrote an analysis program (FIG. 6-10), to calculate the summary statistics for each team. These statistics are percentages of games won at home and away, as well as the average points earned at home and away. Figure 6-11 illustrates the summary statistics for soccer and FIG. 6-12 shows an example of the results of the analysis for basketball. Figure 6-13 summarizes the project procedures.

```
        DATA BASE QUERIES - SOCCER


.USE SOCCER
.
.COUNT FOR HOME_WIN="Y"
        32 records
.
.COUNT FOR HOME_WIN="N"
        17 records
.
.AVERAGE HOME_SCORE
        49 records averaged
        HOME_SCORE
           4.29
.
.AVERAGE VIS_SCORE
        49 records averaged
        VIS_SCORE
           3.57
.
.AVERAGE WIN_SCORE
        49 records averaged
        WIN_SCORE
           5.08
.
.?(4.29+3.57)/2
           3.94
```

6-6 Database query, soccer.

When the calculations were complete, Matthew transferred his results to Microsoft Works on a Macintosh computer. He created pie charts showing the percentage of home games won and lost for both soccer and basketball. He also created bar graphs, which illustrate the average home and visiting scores. Matthew illustrated the MISL results through FIG. 6-14, a pie chart, and FIG. 6-15, a bar graph. He displayed the NBA statistics by again using a pie chart and a bar graph, FIGS. 6-16 and 6-17.

Matthew summarized the results and drew his conclusion, as shown in FIG. 6-18. The results clearly show that statistically, there is a home field advantage, which proves his hypothesis.

On the backboard, Matthew used his graphs, charts, tables, and printed material, as well as several sports photographs, to create an attractive display (FIG. 6-19). Matthew enjoyed doing his project because it dealt with sports and it proved his hypothesis.

SOCCER DATA

HOME_TEAM	VIS_TEAM	HOME_SCORE	VIS_SCORE	HOME_WIN	WIN_SCORE	DATE
STL	KAN	2	3	N	3	10/27/89
DAL	SAN	9	3	Y	9	10/28/89
KAN	STL	6	5	Y	6	10/28/89
WIC	TAC	4	3	Y	4	10/28/89
BAL	CLE	6	3	Y	6	10/29/89
KAN	TAC	5	4	Y	5	10/31/89
CLE	WIC	4	5	N	5	11/02/89
BAL	WIC	7	6	Y	7	11/04/89
DAL	STL	5	4	Y	5	11/04/89
TAC	SAN	4	3	Y	4	11/04/89
TAC	CLE	5	1	Y	5	11/03/89
SAN	KAN	4	3	Y	4	11/03/89
BAL	TAC	3	2	Y	3	11/09/89
CLE	TAC	5	0	Y	5	11/10/89
STL	WIC	2	5	N	5	11/10/89
WIC	BAL	3	9	N	9	11/11/89
DAL	KAN	5	4	Y	5	11/11/89
STL	CLE	4	3	Y	4	11/12/89
SAN	DAL	6	2	Y	6	11/12/89
WIC	DAL	0	3	N	3	11/17/89
SAN	STL	7	4	Y	7	11/17/89
SAN	TAC	4	3	Y	4	11/18/89
CLE	WIC	10	4	Y	10	11/18/89
BAL	KAN	4	2	Y	4	11/18/89
DAL	STL	3	4	N	4	11/19/89
BAL	DAL	3	7	N	7	11/21/89
KAN	CLE	3	4	N	4	11/22/89
WIC	SAN	4	2	Y	4	11/22/89
STL	TAC	2	5	N	5	11/22/89
CLE	DAL	8	3	Y	8	11/24/89
KAN	SAN	4	3	Y	4	11/24/89
WIC	KAN	8	2	Y	8	11/25/89
SAN	BAL	4	5	N	5	11/25/89
TAC	STL	4	6	N	6	11/25/89
DAL	BAL	3	2	Y	3	11/26/89
SAN	KAN	5	4	Y	5	11/28/89
CLE	BAL	4	3	Y	4	12/01/89
STL	SAN	0	4	N	4	12/01/89
WIC	DAL	4	3	Y	4	12/01/89
BAL	CLE	2	3	N	3	12/02/89
DAL	KAN	6	4	Y	6	12/02/89
SAN	WIC	3	5	N	5	12/02/89
TAC	STL	4	5	N	5	12/02/89
KAN	DAL	3	1	Y	3	12/08/89
STL	BAL	2	5	N	5	12/08/89
TAC	WIC	6	5	Y	6	12/08/89
BAL	STL	5	1	Y	5	12/09/89
DAL	SAN	4	1	Y	4	12/09/89
TAC	CLE	2	4	N	4	12/09/89
		4.29	3.57		5.08	
		2.01	1.64		1.64	

6-7 Soccer data listing and averages.

```
DATA BASE QUERIES - BASKETBALL

.USE BBALL
.
.COUNT FOR HOME_WIN="Y"
      163 records
.
.COUNT FOR HOME_WIN="N"
      67 records

.AVERAGE HOME_SCORE
      230 records averaged
      HOME_SCORE
          108.80
.
.AVERAGE VIS_SCORE
      230 records averaged
      VIS_SCORE
          102.15
.
.AVERAGE WIN_SCORE
      230 records averaged
      WIN_SCORE
          115.29
.
.?(108.8+102.15)/2
          105.48
```

6-8 Database query, basketball.

He also learned a lot about computers, and became quite expert at using dBase IV, a complex database program.

Matthew intends to expand on the project next year by including different sports—such as football, baseball, and hockey—and by using a greater portion of the sports season.

THAT'S THE WAY THE BALL BOUNCES

Another interesting project about sports arose from a dinnertime conversation that Aaron Barclay had with his family. How high does a tennis ball bounce? Does it bounce higher on a hard court, on clay, or on grass? An engineering project that measured the bounce of tennis balls launched from the same height at equal velocity determined that a hard court creates the highest bounce.

BASKETBALL DATA

HOME_TEAM	VIS_TEAM	HOME_SCORE	VIS_SCORE	HOME_WIN	WIN_SCORE	DATE
BOS	CLE	102	89	Y	102	12/01/89
NJ	MIA	101	77	Y	101	12/01/89
IND	ORL	125	110	Y	125	12/01/89
WAS	PHI	107	90	Y	107	12/01/89
ATL	UTA	114	103	Y	114	12/01/89
PHO	LAC	111	90	Y	111	12/01/89
LAK	DET	97	108	N	108	12/01/89
POR	GOL	123	110	Y	123	12/01/89
WAS	UTA	98	100	N	100	11/02/89
ATL	PHI	100	92	Y	100	12/02/89
CLE	MIN	74	101	N	101	12/02/89
HOU	DAL	103	106	N	106	12/02/89
SAN	CHA	118	110	Y	118	12/02/89
DEN	POR	146	113	Y	146	12/02/89
PHO	NY	112	122	N	122	12/02/89
SEA	DET	120	95	Y	120	12/02/89
LAC	SAC	114	84	Y	114	12/02/89
GOL	MIL	101	98	Y	101	12/02/89
LAK	NY	115	104	Y	115	12/03/89
ORL	POR	121	95	N	121	12/04/89
NY	PHI	110	103	Y	110	12/05/89
CHA	BOS	101	114	N	114	12/05/89
MIA	POR	107	113	N	113	12/05/89
CLE	UTA	80	96	N	96	12/05/89
CHI	DEN	119	99	Y	119	12/05/89
MIN	NJ	92	90	Y	92	12/05/89
DAL	GOL	107	88	Y	107	12/05/89
SEA	HOU	133	123	Y	133	12/05/89
LAK	LAC	111	103	Y	111	12/05/89
SAC	MIL	118	103	Y	118	12/05/89
BOS	NY	113	98	Y	113	12/06/89
PHI	MIA	121	98	Y	121	12/06/89
ORL	ATL	110	118	N	118	12/06/89
DET	WAS	115	107	Y	115	12/06/89
IND	DEN	136	117	Y	136	12/06/89
SAN	GOL	121	119	Y	121	12/06/89
CHA	POR	86	96	N	96	12/07/89
UTA	DAL	107	97	Y	107	12/07/89
LAK	PHO	100	96	Y	100	12/07/89
LAC	CLE	105	88	Y	105	12/07/89
BOS	DEN	103	102	Y	103	12/08/89
PHI	DET	107	101	Y	107	12/08/89
MIA	ORL	122	114	Y	122	12/08/89
ATL	POR	127	120	Y	127	12/08/89
IND	CHI	106	104	Y	106	12/08/89
HOU	NJ	94	99	N	99	12/08/89
DAL	SAN	93	99	N	99	12/08/89
PHO	MIL	123	98	Y	123	12/08/89
GOL	SAC	121	126	N	126	12/08/89
NY	BOS	124	92	Y	124	12/09/89
WAS	LAK	103	101	Y	103	12/09/89
CHA	DEN	93	106	N	106	12/09/89
ATL	MIN	104	91	Y	104	12/09/89
DET	IND	121	93	Y	121	12/09/89
CHI	PHI	125	105	Y	125	12/09/89
SAN	NJ	109	92	Y	109	12/09/89
UTA	HOU	104	98	Y	104	12/09/89

BASKETBALL DATA

HOME_TEAM	VIS_TEAM	HOME_SCORE	VIS_SCORE	HOME_WIN	WIN_SCORE	DATE
SEA	LAC	104	100	Y	104	12/09/
LAC	CLE	101	108	N	108	12/09/
		108.80	102.15		115.29	
		13.90	11.80		56.50	

6-9 Basketball data listing and averages.

```
ANAL.PRG 12/29/89

STOP = "F"
DO WHILE STOP = "F"
  CLEAR
  *TEAM = "A TEAM'S NAME"
  *LNAME = "FULL TEAM NAME"
  ACCEPT "ENTER TEAM NAME: " TO TEAM
  ACCEPT "ENTER FULL NAME FOR TEAM: " TO LNAME
  HCNT = 0
  VCNT = 0
  HPTS = 0
  VPTS = 0
  HAVE = 0
  VAVE = 0
  HPER = 0
  VPER = 0
  HWIN = 0
  VWIN = 0
  HAVPTS = 0
  VAVPTS = 0
  HPERWIN = 0
  VPERWIN = 0
  GO TOP
  DO WHILE .NOT. EOF()
    IF HOME_TEAM = TEAM
      HCNT = HCNT + 1
      HPTS = HPTS + HOME_SCORE
      IF HOME_WIN = "Y"
        HWIN = HWIN + 1
      ENDIF
    ENDIF
    IF VIS_TEAM = TEAM
      VCNT = VCNT + 1
      VPTS = VPTS + VIS_SCORE
      IF HOME_WIN = "N"
        VWIN = VWIN + 1
      ENDIF
    ENDIF
    SKIP
  ENDDO
  HAVPTS = HPTS / HCNT
  VAVPTS = VPTS / VCNT
  HPERWIN = HWIN / HCNT
  VPERWIN = VWIN / VCNT
  CLEAR
  SET PRINT ON
  ?LNAME
  ?" "
  ?"PERCENT WON AT HOME :    ",HPERWIN
  ?"PERCENT WON VISITING :   ",VPERWIN
  ?"AVERAGE PTS AT HOME :    ",HAVPTS
  ?"AVERAGE PTS AS VISITOR :",VAVPTS
  ?" "
  ?" "
  SET PRINT OFF
  ACCEPT "I WANT TO STOP (T/F):" TO STOP
ENDDO
```

6-10 Analysis program.

SAN DIEGO SOCKERS

```
PERCENT WON AT HOME :        0.71
PERCENT WON VISITING :       0.17
AVERAGE PTS AT HOME :        4.71
AVERAGE PTS AS VISITOR :     2.67
```

CLEVELAND CRUNCH

```
PERCENT WON AT HOME :        0.80
PERCENT WON VISITING :       0.50
AVERAGE PTS AT HOME :        6.20
AVERAGE PTS AS VISITOR :     3
```

KANSAS CITY COMETS

```
PERCENT WON AT HOME :        0.80
PERCENT WON VISITING :       0.14
AVERAGE PTS AT HOME :        4.20
AVERAGE PTS AS VISITOR :     3.14
```

TACOMA STARS

```
PERCENT WON AT HOME :        0.50
PERCENT WON VISITING :       0.17
AVERAGE PTS AT HOME :        4.17
AVERAGE PTS AS VISITOR :     2.83
```

ST. LOUIS STEAMERS

```
PERCENT WON AT HOME :        0.17
PERCENT WON VISITING :       0.43
AVERAGE PTS AT HOME :        2
AVERAGE PTS AS VISITOR :     4.14
```

DALLAS SIDEKICKS

```
PERCENT WON AT HOME :        0.86
PERCENT WON VISITING :       0.33
AVERAGE PTS AT HOME :        5
AVERAGE PTS AS VISITOR :     3.17
```

6-11 Analyzed data, MISL.

LOS ANGELES LAKERS

```
PERCENT WON AT HOME :            0.91
PERCENT WON VISITING :           0.57
AVERAGE PTS AT HOME :          112.27
AVERAGE PTS AS VISITOR :       101.71
```

PORTLAND TRAILBLAZERS

```
PERCENT WON AT HOME :            0.91
PERCENT WON VISITING :           0.50
AVERAGE PTS AT HOME :          113.18
AVERAGE PTS AS VISITOR :       103.63
```

SEATTLE SUPERSONICS

```
PERCENT WON AT HOME :            0.89
PERCENT WON VISITING :           0.25
AVERAGE PTS AT HOME :          117.22
AVERAGE PTS AS VISITOR :       112.88
```

PHOENIX SUNS

```
PERCENT WON AT HOME :            0.70
PERCENT WON VISITING :           0
AVERAGE PTS AT HOME :          115.30
AVERAGE PTS AS VISITOR :       108.40
```

GOLDEN STATE WARRIORS

```
PERCENT WON AT HOME :            0.44
PERCENT WON VISITING :           0
AVERAGE PTS AT HOME :          110
AVERAGE PTS AS VISITOR :       104.33
```

LOS ANGELES CLIPPERS

```
PERCENT WON AT HOME :            0.67
PERCENT WON VISITING :           0
AVERAGE PTS AT HOME :          107.67
AVERAGE PTS AS VISITOR :        97.67
```

6-12 Analyzed data, NBA.

Procedures
1. Built database files to support information requirements.
2. Collected daily scores from all NBA and MISL games from 10/27/89 through 12/09/89.
3. Entered scores into dBase IV format.
4. At end of 1/5 season:
a. Designed and printed database reports to show data collected.
b. Queried database to show and tabulate summary data for:
1) Home score.
2) Visiting score.
3) Winning score.
c. Designed computer program to analyze data
d. Created graphs to reflect summarized and analyzed data.

6-13 Procedures for *Home Field Advantage*.

Like Matthew's project, Aaron took advantage of the latest high-tech gear. He conducted the entire experiment in a little over an hour, while his father videotaped the results! The backboard for the project, entitled Court Surface Effect on Tennis Balls, is shown in FIG. 6-20.

Whatever you're into, for fitness or for fun, you can find a science project.

HELPFUL HINTS

☐ If you're interested in sports, you can find ideas in many categories, physics, engineering, mathematics, or even zoology.

☐ Take advantage of any technology that is available to you, but you do not need a computer, a video camera or other sophisticated equipment to conduct an experiment and do a good project.

SOCCER
GAMES WON BY HOME TEAM

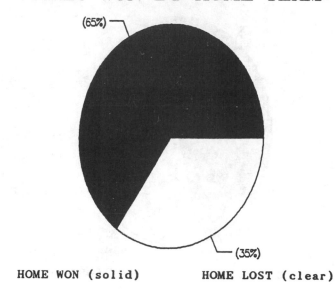

HOME WON (solid)   **HOME LOST (clear)**

6-14 Pie chart, MISL.

SOCCER
AVERAGE SCORES

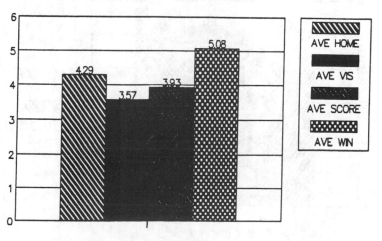

6-15 Bar graph, MISL.

BASKETBALL
GAMES WON BY HOME TEAM

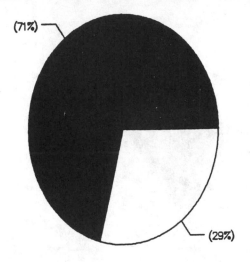

(71%)

(29%)

HOME WON (solid) **HOME LOST (clear)**

6-16 Pie chart, NBA.

BASKETBALL
AVERAGE SCORES

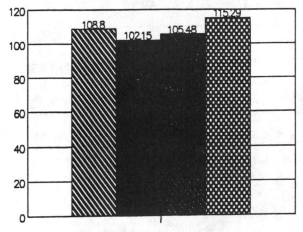

6-17 Bar graph, NBA.

```
|==================================================================|
|                     SUMMARY OF RESULTS                           |
|------------------------------------------------------------------|
|                              SOCCER        BASKETBALL            |
|==================================================================|
|                                                                  |
|  AVG.  HOME  SCORE            4.29           108.8                |
|                                                                  |
|  AVG.  VISITORS  SCORE        3.57           102.15               |
|                                                                  |
|  AVG.  TEAM  SCORE            3.93           105.48               |
|                                                                  |
|  AVG.  WINNING  SCORE         5.08           115.29               |
|                                                                  |
|  PCT.  WON  BY  HOME  TEAM    65%            71%                  |
|                                                                  |
|  TEAMS  WIN  MORE  AT  HOME   7 OF 8         27  OF  27           |
|                                                                  |
|  TEAMS  SCORE  BETTER  AT  HOME   5 OF 8     22  OF  27           |
|==================================================================|
```

The data and resulting analysis indicate that:

1. Home teams win more games.

2. Home teams scores more points.

3. The average home score is greater than the average visitors score.

4. Average visitors score less than average score.

5. Majority of teams win more games at home.

6. Majority of teams score better at home.

6-18 Results and conclusion, *Home Field Advantage*.

6-19   Backboard, *Home Field Advantage*.

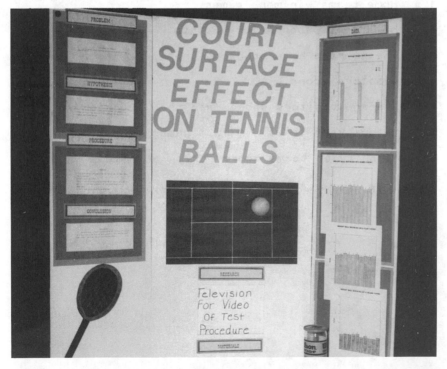

6-20   Backboard display; *Effect of Conditions on the Bounce of a Tennis Ball*.

Chapter **7**

Science project organizer

Now it's time to get started on your own science project. As you have seen, the first and often the hardest step is to come up with an idea. You might start with some of your interests or hobbies, an item you've heard or read about, or something you're studying in class. However, one of the best ways to get going is to brainstorm, either alone, with your family, with friends, or with your teacher. Just think of as many things as you can, and write them down. To keep track of your project ideas, use the handy-dandy list shown in FIG. 7-1.

Next, select the idea you think is the most promising, and do a little reality checking by reviewing the project idea checklist in FIG. 7-2. If, for any reason, you think that your original idea just won't fly, go to your next favorite idea, and run through the checklist again. Even after you think you've found the perfect idea, discuss it with your teacher. He or she may know something you don't—that it's too easy, too complex, might be disqualified, or that half the city used that same topic last year!

Another thing you may want to do before making a final decision on an idea is to see what resources are available, both for your background research paper, and for any help you may need while doing the project. Daryl Smith used this as a crucial step before making his decision on a project idea. Figure 7-3 will help you check out the possible sources of information and help.

PROJECT IDEAS

☻

☻

☻

☻

☻

☻

☻

☻

☻

☻

☻

7-1   Project ideas

PROJECT IDEA CHECKLIST

√ Is the information I need readily available to me?

√ Where can I find the information?

√ If the information is not available locally, where is it? How long will it take to get?

√ Will I need to pay for the information (for example, government pamphlets?

√ If I need special books, can I check them out of the library or will I need to use them there?

√ Will I need professional advice? From whom? Are they willing to help?

√ What materials will I need? What will they cost?

√ Can I borrow some of the supplies and equipment I need?

√ Can I build some of the things I need? Do I need help?

√ Can I finish this project in the time available? If not, can I take a small portion of it to work on?

√ Is there anything about this project my family will object to?

7-2 Project idea checklist

SOURCES OF INFORMATION AND HELP

- Libraries

    ☻

    ☻

    ☻

- Research facilities

    ☻

    ☻

    ☻

- Universities

    ☻

    ☻

    ☻

- Government publications

    ☻

    ☻

    ☻

- Businesses

    ☻

    ☻

    ☻

- Interviews

    ☻

    ☻

    ☻

7-3 Sources of information

Now that you've decided and perhaps done some background research, you'll want to get down to the details. Use FIG. 7-4 to state your question and hypothesis. Be as specific as possible when stating your hypothesis because this is what your project must prove or disprove.

Figures 7-5 and 7-6 will help you outline your variables and controls, as well as your experimental and control groups, if those elements are part of your project. As you've seen, not all projects will use variables, controls, experimental groups or control groups. You may also want to use the table in FIG. 7-7 to outline your sample size or the number of trials you will do.

Incidentally, don't get carried away with all the forms. They're just here to remind you of some of the elements of a science project.

With any project, it's important to have everything you need before you start. Use the table in FIG. 7-8 to make up your project shopping list, and perhaps your project budget as well.

To keep on track once you actually begin your experiment or program, you'll want to know exactly what you'll be doing and in what sequence. Use the table shown in FIG. 7-9 to carefully and specifically list your procedures.

Don't leave out any steps. Something that seems minor could be an important part of your procedures. Include anything that is part of the start-up procedure before the actual testing. For example, Jennifer Ade had to label and punch holes in 120 cups and plant the radish seeds before she could begin her experiment on acid rain.

Finally, you'll need to make sure that your record keeping is complete and accurate. If you do not already have a schedule or a project calendar, you can use the one shown in FIG. 7-10 to keep you on track, from the day you begin until the science fair.

Besides your schedule, maintain a project log showing what you did and when you did it. This could be extremely helpful later on, especially if your project is late due to factors beyond your control. A sample of a project log is shown in FIG. 7-11.

Your experimental log, however, may take many forms, depending on the nature of your experiment. Brian Berning and Nguyen Vy had to record daily findings in order to have accurate results, while Yolanda Lockhart recorded results as she tested. In any event, if you look back through the projects shown, you'll see that each log is very individual, depending on the nature of the project.

QUESTION & HYPOTHESIS

QUESTION	HYPOTHESIS

QUESTION	HYPOTHESIS

QUESTION	HYPOTHESIS

QUESTION	HYPOTHESIS

7-4 Questions and hypotheses

VARIABLES & CONTROLS

VARIABLES	CONTROLS
Experimental	1.
1.	2.
2.	3.
3.	4.
4.	5.
5.	
Measured	
1.	
2.	
3.	
4.	
5.	

VARIABLES	CONTROLS
Experimental	1.
1.	2.
2.	3.
3.	4.
4.	5.
5.	
Measured	
1.	
2.	
3.	
4.	
5.	

7-5 Variables and controls

EXPERIMENTAL & CONTROL GROUPS

EXPERIMENTAL GROUPS	CONTROL GROUP
1.	
2.	
3.	
4.	
5.	

EXPERIMENTAL GROUPS	CONTROL GROUP
1.	
2.	
3.	
4.	
5.	

EXPERIMENTAL GROUPS	CONTROL GROUP
1.	
2.	
3.	
4.	
5.	

7-6 Experimental and control groups

SAMPLE SIZE & NUMBER OF TRIALS

SAMPLE SIZE	NUMBER OF TRIALS

SAMPLE SIZE	NUMBER OF TRIALS

SAMPLE SIZE	NUMBER OF TRIALS

SAMPLE SIZE	NUMBER OF TRIALS

7-7 Sample size and number of trials

EQUIPMENT & SUPPLIES

Material	Source	Cost

7-8 Equipment and supplies

PROCEDURES LIST

PROCEDURES
1.
2.
3.
4.
5.
6.
7.
8.
9.
10.
11.
12.
13.
14.
15.
16.
17.
18.
19.
20.

7-9 Procedures list

Again, these forms are here to help you, not to restrict or limit you. If your project (or your individual working style) does not lend itself to this type of structure, feel free to take off on your own!

PROJECT SCHEDULE

DATE DUE	ACTIVITY	DATE DONE

7-10 Project schedule

PROJECT LOG

DATE	TIME	ACTIVITY

7-11 Project log

Appendix

ISEF project categories

The following are the project categories officially recognized by the International Science and Engineering Fairs (ISEF).

BEHAVIORAL AND SOCIAL SCIENCES

Psychology, sociology, anthropology, archaeology, ethology, ethnology, linguistics, animal behavior (learned or instinctive), learning, perception, urban problems, reading problems, public opinion surveys, and educational testing, etc.

BIOCHEMISTRY

Molecular biology, molecular genetics, enzymes, photosynthesis, blood chemistry, protein chemistry, food chemistry, hormones, etc.

BOTANY

Agriculture, agronomy, horticulture, forestry, plant biorhythms, palynology, plant anatomy, plant taxonomy, plant physiology, plant pathology, plant genetics, hydroponics, algology, mycology, etc.

CHEMISTRY

Physical chemistry, organic chemistry (other than biochemistry), inorganic chemistry, materials, plastics, fuels, pesticides, metallurgy, soil chemistry, etc.

COMPUTER SCIENCE

New developments in hardware or software, information systems, computer systems organization, computer methodologies and data (including structures, encryption, coding and information theory).

EARTH AND SPACE SCIENCES

Geology, geophysics, physical oceanography, meteorology, atmospheric physics, seismology, petroleum, geography, speleology, mineralogy, topography, optical astronomy, radio astronomy, astrophysics, etc.

ENGINEERING

Civil, mechanical, aeronautical, chemical, electrical, photographic, sound, automotive, marine, heating and refrigerating, transportation, environmental engineering, etc. Power transmission and generation, electronics, communications, architecture, bioengineering, lasers, etc.

ENVIRONMENTAL SCIENCES

Pollution (air, water, land), pollution sources and their control, waste disposal, impact studies, environmental alteration (heat, light, irrigation, erosion, etc.), ecology.

MATHEMATICS

Calculus, geometry, abstract algebra, number theory, statistics, complex analysis, probability, topology, logic, operations research, and other topics in pure and applied mathematics.

MEDICINE AND HEALTH

Medicine, dentistry, pharmacology, veterinary medicine, pathology, ophthalmology, nutrition, sanitation, pediatrics, dermatology, allergies, speech and hearing, optometry, etc.

MICROBIOLOGY

Bacteriology, virology, protozoology, fungal and bacterial genetics, yeast, etc.

PHYSICS

Solid state, optics, acoustics, particle, nuclear, atomic, plasma, superconductivity, fluid and gas dynamics, thermodynamics, semiconductors, magnetism, quantum mechanics, biophysics, etc.

ZOOLOGY

Animal genetics, ornithology, ichthyology, herpetology, entomology, animal ecology, anatomy, paleontology, cellular physiology, animal biorhythms, animal husbandry, cytology, histology, animal physiology, neurophysiology, invertebrate biology, etc.

CATEGORY INTERPRETATIONS

Below are project areas about which questions frequently arise. It is included only to provide some basis for interpretation of the category descriptions.

Instruments

The design and construction of a telescope, bubble chamber, laser, or other instrument would be Engineering if the design and construction were the primary purpose of the project. If a telescope were constructed, data gathered using the telescope, and an analysis presented, the project would be placed in Earth and Space Sciences.

Marine Biology

Behavioral and Social Sciences (schooling of fish), Botany (marine algae), Zoology (sea urchins), or Environmental Sciences (plant and animal life of sea, river, pond).

Fossils

Botany (prehistoric plants), Chemistry (chemical composition of fossil shells), Earth and Space Sciences (geological ages), and Zoology (prehistoric animals).

Rockets

Chemistry (rocket fuels), Earth and Space Sciences (use of a rocket as a vehicle for meteorological instruments), Engineering (design of a rocket), or Physics (computing rocket trajectories), A project on the effects of rocket acceleration on mice would go in Medicine and Health.

Genetics

Biochemistry (studies of DNA), Botany (hybridization), Microbiology (genetics of bacteria), or Zoology (fruit flies).

Vitamins

Biochemistry (how the body deals with vitamins), Chemistry (analysis), and Medicine and Health (effects of vitamin deficiencies).

Crystallography

Chemistry (crystal composition), Mathematics (symmetry), and Physics (lattice structure).

Speech and hearing

Behavioral and Social Sciences (reading problems), Engineering (hearing aids), Medicine and Health (speech defects), Physics (sound), Zoology (structure of the ear).

Radioactivity

Biochemistry, Botany, Medicine and Health, and Zoology could all involve the use of tracers. Earth and Space Sciences or Physics could involve the measurement of radioactivity. Engineering could involve design and construction of detection instruments.

Space related projects

Note that many projects involving "space" do not go into Earth and Space Sciences. Botany (effects of zero G on plants), Medicine and Health (effects of G on human beings), Engineering (development of closed environmental system for space capsule).

Computers

If a computer is used as an instrument, the project should be considered for assignment to the area of basic science on which the project focuses. As examples: if the computer is used to calculate rocket trajectories, then it would be assigned to Physics, or if the computer is used to calculate estimates of heat generated from a specified inorganic chemical reaction, then it would be entered in Chemistry, or if the computer is used as a teaching aid, then it would be entered in Behavioral and Social Sciences.

Glossary

abstract A short summary of the main points of a project. This is normally between 200 and 250 words in length.

analyzed data Data derived from raw and smooth data, from which conclusions can be drawn.

conclusion Interpretation based on outcome of results and answering the question or comparison suggested by purpose.

control group Identical to the experimental group in all aspects, except that no variables are applied. This represents the test group that has all variables standardized and forms the basis for comparison.

controls This represents factors which are not to be changed or variables which are to be controlled. DO NOT confuse with control group.

dependent variable The factor which changes as a result of altering the independent variable. Also, the change in events or results linked and controlled by another factor which has also been changed.

experiment A planned investigation to determine the outcome which would arise from changing a variable or from changing "natural" conditions.

experimental group A group of subjects to which independent or experimental variables are applied.

experimental variable See *independent variable*.

graphs Illustrated form of presenting raw, smooth or analyzed data.

hypothesis   Statement of an idea that can be tested experimentally, based upon research. States what experimenter believes will happen as a result of the experiment.

independent variable   The item, quantity, or condition which is altered to observe what will happen; something that can be changed in an experiment without causing a change in other variables.

interpretation   One's personal viewpoint based on the data. This can be based on either qualitative or quantitative analysis and may become a part of the project's conclusion.

materials   All items used in the course of the experiment.

measured variable   See *dependent variable.*

observation   What one sees in the course of the experiment. Observations are often incorporated into raw data.

procedures   Steps which must be followed to perform an experiment.

proposal   The planned procedure for experimentation.

qualitative analysis   Analysis made subjectively, without measurement.

quantitative analysis   Analysis made objectively, with measurement devices.

question (or problem)   That which forms the basis of the hypothesis and hence is the objective of the experiment.

raw data   Logs, tables, and graphs which represent data as it is collected in the course of the experiment.

research   The process of learning facts or prior theories on a subject by reviewing existing sources of information.

results   Graphs and tables which represent raw, smooth and analyzed data.

scientific method   Manner of conducting an experiment, using valid subjects, variables and controls, and accurately recording results.

smooth data   Tables or graphs where all the averages, totals or percentages are placed. These may combine the information from raw data tables and graphs.

tables   Written form of presenting raw, smooth or analyzed data.

variable   A condition which is changed to test the hypothesis or a condition which changes as a result of testing the hypothesis.

Bibliography

The following bibliographies are organized by project name. Only those projects for which students actually submitted bibliographies are listed.

THE GOOD EARTH   Chapter 2

Brown, Howard E., Victor E. Monnett, and J. Willis Stovall. *Introduction to Geology*. Boston: Ginn and Co., 1958.

The Encyclopedia Americana. "Conservation." 1989. Vol. 10, p. 556.

Foster, Albert E. *Approved Practices in Soil Conservation*. Illinois: The Interstate Inc., 1973.

Goldman, Steven J., Katherine Jackson, and Taras A. Bursztyn *Erosion and Sediment Control*. New York: McGraw-Hill Co., 1986.

Kohnke, Helmut and Anson R. Bertrand. *Soil Conservation*. New York: McGraw-Hill Co., 1959.

Mowitz, Greg. *Successful Farming*. "How Erosion Wrecks and Yields Profits." May, 1983. Vol. 81, pp. 10-11.

Newsweek. "The Disappearing Land." Aug. 23, 1982. Vol. 10, pp. 24-26.

Smith, Grahame J. C., Henry J. Steck, and Gerald Surette. *Our Ecological Crisis*. New York: MacMillan Publishing Co.

Wood, Greg, Mark Pearson, and Cheryl Tevis. *Successful Farming*. "Corn Residue Stops Erosion, Nutrient and Herbicide Loss." May, 1983. Vol. 81, pp. 10-11.

The World Book Encyclopedia. "Conservation." 1987. Vol. 3, p. 353.

IT'S RAINING, IT'S POURING   Chapter 2

Air Pollution Control District County of San Diego, *Air Quality in San Diego*. San Diego: San Diego Air Pollution Control District, 1984.

Asimov, Isaac. *The Noble Gases*. New York: Basic Books Inc., 1966.

Bender, David and Bruno Leone. *The Environmental Crisis Opposing Viewpoints*. Minnesota: Greenhaven Press, 1986.

Bloome, Enid. *The Air We Breathe*. New York: Doubleday & Company, 1971.

Carr, Donald. *The Sky is Still Falling*. New York: W. W. Norton & Company, 1982.

Elliott, Sarah. *Our Dirty Air*. New York: Julian Messner, 1971.

Gould, Roy. *Going Sour*. Boston: Birkhauser, 1985.

Herron, J. Duddley. *Understanding Chemistry, 2nd Edition*. New York: Random House, 1986.

Hess, Fred, revised by Arthur Thomas. *Chemistry Made Simple*. New York: Doubleday & Company, 1984.

Hyde, Margret. *For Pollution Fighters Only*. San Francisco: McGraw-Hill Book Company, 1984.

Jones, Claire, Steve Gadler, and Paul Engstorm. *Pollution: The Air We Breathe*. Minnesota: Lerner Publishing Co., 1971.

King, Jonathon. *Troubled Waters*. Pennsylvania: Rodeo Press.

JOLLY ORVILLE   Chapter 3

Chinnici, Madeline. "Down with U.P. Unpopped Popcorn." *Science World*. March 9, 1990, p. 7.

Redenbacher, Orville. *Orville Redenbacher's Popcorn Book*. New York: St. Martin's Press, 1984.

World Book Encyclopedia. 1981. Volume 15, p. 585.

GREEN THUMB   Chapter 3

Britannica Junior Encyclopedia. 1978. "Plant." Vol. 12, pp. 216−218.

Comptons Encyclopedia and Fact Index. "Plants." 1978. Vol. 12, pp. 92−97.

Halsey, David D. "Plants." *Merits Student's Encyclopedia*. 1986. Vol. 4, pp. 102−121.

Heilmer, Charles H. *Focus on Life Science*. Charles E. Merril Publishing Company; a Bell Howell Company 1984.

New Age Encyclopedia. 1976. Vol. 15, pp. 271−272.

DO YOU SPEAK BASIC   Chapter 4

Denning, Peter J., et al. "Computing as a Discipline." *Communications of the ACM*. 32 (January, 1989), pp. 9–23.

Holt, R. C. and J. R. Cordy. "The Turing Programming Language." *Communications of the ACM*. 31 (December, 1989), pp. 1410–1421.

Horn, L. Wayne, and Michel Boillot. *Basic Fourth Edition*. Los Angeles: West Publishing Company, 1986.

Sammet, Jean E. *Programming Languages: History and Fundamentals*. Englewood Cliffs: Prentice-Hall Inc., 1969.

Tennent, R. D. *Principles of Programming Languages*. Englewood Cliffs: Prentice-Hall Inc., 1981.

Weingarten, Frederick W. *Translation of Computer Languages*. San Francisco: Holden-Day Inc., 1973.

Wexelblat, Richard L. *History of Programming Languages*. New York: Academic Press, 1981.

GW-BASIC Interpreter Version 2. Microsoft. 1984.

Microsoft MS-DOS Version 3 — Basic Concepts and Features. Zenith Data Systems. 1986.

Microsoft MS-DOS Version 3 — Primary Command Guide. Zenith Data Systems. 1986.

WHERE THERE'S SMOKE . . .   Chapter 5

Anthony, Catherine and Norma Kolthoff. *Anatomy and Physiology*. C. V. Moshy Company, 1971.

Asimov, Isaac. *The Human Body*. Boston: Houghton-Mifflin Company, 1963.

Health Consequences of Involuntary Smoking, revised edition. U.S. Department of Health and Human Services, 1986.

Hedrich, James. *Smoking Tobacco and Health*. Maryland: Resource Management Corporation.

Rubenstein and Federman, *Scientific American: Medicine*. New York: Scientific American.

Scientific Cases Against Smoking, revised edition. U.S. Department of Health. Washington, D.C. 1980.

Terry, Luther and Daniel Horn., *To Smoke or Not to Smoke*. New York: Lothrop, Lee and Shepard Co., 1969.

Index

A

acid rain, 18-29
Ade, Jennifer L., plant growth vs.
    drainage, 37-41
animal rights and ethics, 6-7
athletics (*see* sports-related experi-
    ments)

B

backboards and displays, 2-5
ball-bounce experiment, 86-90, 94
bar graphs, 91-92
Barclay, Aaron, ball-bounce experi-
    ment, 86-90, 94
BASIC programming, language trans-
    lator, 47-61
behavioral and social science, 77, 109
    animals or humans as subjects, 6-7
Berning, Brian, erosion experiment,
    9-17, 99
biochemistry, 109
botany, 109
bouncing balls, 86-90, 94
brainstorming, 95-97
budgeting, 99

C

categories of projects, 109-112
chemistry, 109
cigarettes, second-hand smoke
    effects, 69-77
computer sciences, 47-62, 110, 112
    home-field advantage, 79-93
conservation (*see* ecology and con-
    servation)
control and experimental groups, 77,
    102
controls and variables, 101
crystallography, 112

D

Davis, Catherine, acid rain experi-
    ment, 9, 18-29
displays and backboards, 2-5

Duke, Matthew, home-field advan-
    tage, 79-93

E

Earth and space sciences (*see* ecol-
    ogy and conservation)
ecology and conservation, 9-29, 110
    erosion and its effects, 10-17
    plant growth vs. drainage, 37-41
engineering, 110
environmental sciences (*see* ecology
    and conservation)
equipment and supplies, 90, 104
erosion and its effects, 10-17
experiment log (*see* logbooks)
experimental and control groups, 77,
    102
experiments, 25, 76

F

foreign language translation pro-
    gram, 47-61
fossils, 111

G

games (*see* sports-related experi-
    ments)
genetics, 111

H

home-field advantage, 79-93
Houlton, David, BASIC program-
    ming, 47-61
household experiments, 31-45
    plant growth vs. drainage, 37-41
    popcorn pop-rate experiment, 31-
    37
    rust prevention, 41-45
human subjects in experiments, 7
hypothesis, 76, 100

I

instruments, scientific, build and
    operate, 111

ISEF project categories, 109-112

L

Lockhart, Yolanda, peripheral vision, 63-69, 99
logbooks, 25, 99, 107

M

marine biology, 111
materials needed, 99
medical sciences, 63-77, 110
   animal experiments, 6-7
   behavioral or social science projects, 77
   human subjects, 7
   peripheral vision, 63-69
   second-hand cigarette smoke, 69-77
microbiology, 110

N

Nguyen, Vy, popcorn experiment, 31-37, 99

P

peripheral vision experiment, 63-69
physics, 110
pie charts, 91-92
plant growth vs. drainage, 37-41
popcorn pop-rate experiment, 31-37
procedures list, 105
programming (*see* computer sciences)
project idea checklist, 97
project log (*see* logbooks)

R

radioactivity, 112
record keeping, 76, 99
research, 25, 76, 98-99
resource materials available, 25, 98-99
rockets, 111
rust prevention, 41-45

S

sample size, 25, 103
schedule for project, 106
science project organizer, 95-107
   equipment and supplies, 104
   experimental and control groups, 102
   logbooks, 99, 107
   materials need and budgeting, 99

   procedures list, 105
   project idea checklist, 97
   project schedule, 106
   question and hypothesis, 100
   record keeping, 99
   resource material available, 98-99
   sample size and number of trials, 103
   topic selection and brainstorming, 95-97
   variables and controls, 101
scientific instruments, build and operate, 111
second-hand cigarette smoke experiment, 69-77
Slater, LeMar, rust prevention, 41-45
Smith, Daryl, second-hand cigarette smoke, 69-77, 95
smoking, second-hand cigarette smoke, 69-77
social sciences, 77, 109
space-related projects, 112
speech and hearing, 112
sports-related experiments, 79-94
   bouncing balls, 86-90, 94
   home-field advantage, 79-93
   physics, engineering, mathematics projects, 90

T

topic selection, 1-7, 95-97
translation programming in BASIC, 47-61
trials, sample sizes, 103

V

variables and controls, 101
vision, peripheral vision experiment, 63-69
vitamins, 112

W

water
   acid rain, 18-29
   erosion and its effects, 10-17
   plant growth vs. drainage, 37-41
   rust prevention, 41-45

X

XLATE language translation program, 47-61

Z

zoology, 111